ECOLOGICAL APPROACH TO
PEST MANAGEMENT

Food Production and Natural Resources
David Pimentel, *Editor*

Ecological Approach to Pest Management
DAVID J. HORN

ECOLOGICAL APPROACH TO
PEST MANAGEMENT

DAVID J. HORN
The Ohio State University

Foreword by David Pimentel

THE GUILFORD PRESS
New York   London

© 1988 The Guilford Press
A Division of Guilford Publications, Inc.
72 Spring Street, New York, N.Y. 10012

Printed in the United States of America

Last digit is print number: 9 8 7 6 5 4 3 2 1

Library of Congress Cataloging-in-Publication Data

Horn, David J.
  Ecological approach to pest management.

  (Food production and natural resources)
  Bibliography: p.
  Includes index.
  I. Insect pests—Control.   2. Insects—Ecology.
I. Title.   II. Series
SB931.H595   1988   632′.7   87-12944
ISBN 0-89862-302-2 (cloth)
ISBN 0-89862-505-X (paper)

FOREWORD

The Food Production and Natural Resources Series has as its broad objective to explore the interdependencies of food production and environmental resources. With a world population of 5 billion people and an additional 270,000 people being added each day, the earth's resources are having to be used to produce the increased amounts of food needed to support life. Now, and even more so in the future, our scientists will have to do a better job to develop new technologies and approaches that not only conserve resources but make our land, water, energy, and biological resources more productive.

The books in the Food Production and Natural Resources Series are designed to be college texts and to serve as references for scientists concerned with the use and management of natural resources for food production. The series of books makes a special effort to cover the basic scientific principles and especially to describe the new technologies that are available for resource management and food production.

One important facet of food production is the control of pests that attack, destroy, and otherwise damage crops, making them unsuitable for human food and livestock forage. Each year in the United States, despite the use of all pesticides and other controls, pests destroy about 37% of all potential crops.

One cost-effective way to substantially increase food supplies while decreasing land and water resource use is to reduce losses caused by pest insects, plant pathogens, and weeds. All too often in the past, pest control has relied on a chemical, *ad hoc* approach and has not taken into consideration an understanding of the ecology of the pests. This approach has resulted in more pesticides being used with greater costs for the farmer. When used routinely, pesticides have damaged beneficial insects and upset the natural controls that exist in agroecosystems.

Thus, newer, more broad-based insect control programs must be developed and employed. In this book, Dr. David Horn has successfully integrated basic ecological principles into workable pest management strategies. Explained are the complexities of pest population dynamics and the various causes of pest outbreaks. As important are the discussions about what determines the abundance and distribution of pests. With this knowledge, scientists are able within reason to alter the agroecosystem to make it unfavorable for pest populations.

Because of his vast personal involvement in ecologically based pest control, Dr. Horn is able to show how to apply ecological knowledge to the development of pest control measures. For example, he illustrates how biological controls can be integrated with various other nonchemical as well as chemical controls.

Dr. Horn points out that the foundation of integrated pest management (IPM) is based on the ecological principle of manipulating agroecosystems to make them unfavorable for pests. Most IPM specialists are now seeking basic training in ecology. Dr. Horn has been a leader in the development of new concepts and ecological approaches in pest control. In this book, he explains his perspective and the soundness of integrating ecology and pest control.

In the final analysis, the truly effective pest control measures are those that are long lasting, economical, and that minimize damage to natural enemies and the environment. Dr. Horn's thorough analysis of all aspects of pest control clarifies both the scientific principles as well as the practical aspects that will enable us to reduce insect pest losses and make agriculture more productive.

David Pimentel
Cornell University

PREFACE

The past decade has seen growing integration between research in basic insect ecology and development of insect pest management. The successful management of any insect pest population today depends upon an increasingly sophisticated knowledge of the functional role of insects in ecosystems. A generation ago, a quick solution to any pest problem was simply a matter of chemical pesticide application. The heyday of exclusive reliance on chemical insecticides for insect control is past, although chemicals remain a mainstay of many integrated approaches to pest management. In a sense, entomologists today have returned to pest management approaches used before the widespread application of synthetic organic biocides, although modern pest management is undergirded with rigorous ecological studies in a way never envisioned or even possible in years past.

This book is designed as a single source addressing principles of insect ecology and their relation to insect pest management. The book has grown out of a course that I have taught for 15 years to advanced undergraduate and graduate students majoring primarily in entomology but also in agronomy, horticulture, natural resources, plant pathology, or zoology. A serious student in any one of these fields should be able to understand the material in this book, though it is intended primarily as a text for courses in insect pest management, which is in turn only a single phase of total environmental management. I have written this integration of insect ecology and pest management in the hope of reaching more than 16 students annually. I have assumed some familiarity with the basic biology of insects, and introductory courses in statistics and ecology are also helpful but not essential to understanding the material herein.

Agricultural examples are emphasized, perhaps reflecting my own interests and background. There is additional material concerning management of forest insects and pests of public health significance. I hope that coverage of the literature and inclusion of personal observations are balanced, but inevitably the examples are selective. As is also inevitable in any rapidly developing field, new advances appear daily and parts of this book will be out of date in short order. (Most of the literature review was completed during 1985.)

I have not tried to cover comprehensively the entire field of insect control; it is too large and I am too ignorant. Examples are included primarily to illuminate major points in the text. Therefore this book is not a detailed guide on how to manage insect pests. There are on the market several excellent books covering the specific details of insect management (see Chapter 12), and current information can be obtained from the U.S. Cooperative Extension Service.

The following persons have read and commented upon portions of this book during its lengthy and sometimes tortuous development: David Pimentel, James R. Carey, Susan W. Fisher, Franklin R. Hall, Richard W. Hall, Ronald B. Hammond, Mark Hertlein, Rosalind S. Horn, J. Mark Scriber, John A. Shuey, Charles G. Summers, and Harold R. Willson. All provided useful comments that improved the manuscript substantially. The students who were enrolled in Entomology 660 at The Ohio State University from 1983 to 1986 provided student-eye views. The comments of two anonymous reviewers were most helpful. Errors doubtless remain, and I will take responsibility for them.

L. E. Ehler and the rest of the faculty, and graduate students, of the Department of Entomology, University of California, Davis, provided a congenial atmosphere in which this project was well-launched. My colleagues in the Department of Entomology, The Ohio State University, have supported me to its completion.

I thank Rose Horn for providing technical assistance in production of the manuscript, and Kathy Horn for modeling sampling techniques (Figures 3-5 and 3-7). Each of these women has encouraged me to maintain my sense of humor during dark times.

Several publishers graciously gave permission for the use of copyrighted material, for which I am grateful. They are acknowledged in the appropriate table and figure captions.

Seymour Weingarten, Editor-in-Chief of The Guilford Press, and David Pimentel, Editor of the Food Production and Natural Resources Series, merit my special thanks for their unwavering interest and support, and their patience regarding deadlines.

I could not have done this or much else without the enduring patience and love of Rosalind Horn, who has cheerfully endured the emotional burden of being married to an author.

This book is dedicated to the memory of two fellow entomologists—my good friends Warren R. Cothran and Dwight M. DeLong.

David J. Horn

CONTENTS

ECOLOGICAL APPROACH TO
PEST MANAGEMENT

CHAPTER

1

INTRODUCTION

BACKGROUND

Since the beginning of recorded history, and probably for much longer than that, people have faced competition from, and direct attack by, insects and related arthropods. For millenia, responses to this challenge were barely adequate, while crop-destroying pests and arthropod-borne diseases wreaked unpredictable though regular havoc on humans and their agriculture. As examples one need only recall locust plagues of antiquity in the Mediterranean region, or the Black Death, the flea-transmitted bubonic plague that killed one of every four Europeans in the mid-14th century. Until World War I, deaths due to louse-transmitted typhus exceeded battle deaths among troops engaged in western European wars (Zinsser, 1938). Gradually, mostly through trial and error, more sophisticated techniques were developed for managing pests, increasing the degree of control and lengthening the intervals between episodes of devastation. A century ago, some insect pests were controlled at acceptable levels much of the time, though unmanageable outbreaks of such pests as locusts, mosquitoes, and others still occurred with distressing frequency. Entire regions of the earth remained susceptible to agricultural ruin following introduction of a new pest, an example being the near-collapse of cotton production in the southeastern United States following the establishment of the boll weevil, spreading northward and eastward from its ancestral home in central Mexico.

A century ago, levels of control prevalent between pest outbreaks usually left a residual pest population behind, a focus for the spread of future infestations and perhaps in numbers still capable of damage locally. In those times, "acceptable management" was not necessarily synonymous with "desirable control."

1

Greater effort to devise more effective schemes for managing insect pests came with the recognition of agricultural science as a legitimate academic specialty (approximately coincident with the 1862 Morrill Act establishing land-grant universities in the United States) and the dawning recognition of the role of arthropods in transmission of pathogenic microbes (during the last quarter of the 19th century). Farmers were encouraged to adjust tillage practices, to apply chemical poisons, and to encourage and conserve natural enemies whenever possible. Increased emphasis on sanitation reduced the incidence of arthropod-borne disease and removed many plant pests that overwintered in crop residue and other waste places. Most of these procedures amounted to what is now called integrated pest management, although growers and farm advisers lacked the sophisticated understanding of ecosystem functioning that undergirds today's management schemes. However, monitoring of insect populations to estimate an economic threshold for chemical treatment has been practiced for over 80 years, beginning with Arkansas cotton in 1901 (Isely, 1942).

Application of chemical poisons is an agricultural practice well over a century old, though adoption of insecticidal control was limited by the paucity of chemicals available and their limited utility, and use of pesticides was a minimal component of agricultural operations. At first, growers were unwilling to accept readily the newfangled idea of chemical agriculture, though glowing optimism about a bug-free future due to chemicals began to be expressed in agricultural magazines as early as 1903. Increasing reliance on chemicals in many phases of agriculture was part of the growing momentum toward mechanization that proceeded at an accelerated pace after the First World War. This trend led farmers and their governmental and industrial support systems to seek greater efficiency to insect management. Initially, choice of chemicals was limited to either inorganic salts (e.g., lead arsenate or Paris green) or organic chemicals derived from plants (e.g., nicotine, pyrethrum, or rotenone). The latter, which are expensive today, were expensive even at Depression-era prices, though they brought unprecedented reductions in insect pest populations. Interest in insecticide use ran high among agricultural and medical entomologists and among growers, well before the advent of DDT (Smith, 1969).

The introduction of synthetic organic insecticides, commencing with DDT, revolutionized insect pest control. The first practical uses of DDT were against the Colorado potato beetle in Switzerland in 1941, and against body lice among Allied troops and civilians in Italy during 1943. The latter use ended a threat of typhus in an active combat area and did much to hasten the adoption of DDT as a "miracle" control for all species of arthropod pests. Starting in 1945 the hope was raised that inexpensive, efficient contact poisons would permit control of insect populations at densities lower than ever before imagined. So powerful did these chemicals seem that some entomologists (including the President of the Entomological Society of America) went so far as to propose conservation to preserve mosquitoes, houseflies, cockroaches, and lice from extinction. Near-universal dependence on chemical insecticides fit easily into the trend toward ever more energy-intensive, highly mechanized agriculture as practiced in the economically developed nations of Europe and North America. The two subsequent decades witnessed vastly increasing reliance on chemical poisons for insect pest control, with abandonment of many alternative techniques used in earlier years.

Even as the rush to develop newer and more effective insecticides continued, some entomologists predicted difficult times ahead for chemical control. Resistance to insecticides was documented as early as 1908 (Melander, 1914), though there were very few instances of resistance before the 1940s. However, by 1948, resistance to DDT had been demonstrated among houseflies and several other medically significant pests (Brown, 1958). Since then, the number of insect species resistant to insecticides has increased to over 430 (Ball, 1981; Dover & Croft, 1986). Heavy and repeated insecticide use also generated secondary pest problems; insects such as bollworms (*Heliothis* spp.) on cotton, and leafrollers on apples, peaches, and other fruits increased in numbers as their predators were eliminated by insecticides applied against boll weevils (on cotton) or codling moths (on apples).

Equally alarming was a growing awareness that extremely stable chemicals—notably, some of the chlorinated hydrocarbons and their metabolic derivatives—were gradually accumulating in the global ecosystem, and that several species of fish and birds were adversely affected by minute amounts of these poisons. Though it was not (and has yet to be) documented that similar tiny quantities of these chemicals in human food have any discernible negative effect, the mere notion that these chemicals were present was most unsettling (Carson, 1962). Some of the shorter-residual compounds—particularly, the organophosphates—were (and still are) extremely hazardous in small amounts. (The more potent organophosphates were first developed during World War II by the Third Reich, for purposes far more sinister than insect pest management.) Heavy use of organophosphates led to an increase in accidental poisonings. The number of accidental deaths due to insecticide poisoning in the United States has never been very great and averages about 30–50 annually (Ware, 1983), though as many as 40,000 nonfatal poisonings are recorded annually (Pimentel & Levitan, 1986). However, the worldwide estimate of 20,000 annual deaths due to pesticide poisoning (Metcalf, 1980) is of very serious concern. Moreover, there is always the possibility of a tragic industrial accident such as the chemical leak in Bhopal, India, in 1984, or of contamination such as the aldicarb-tainted watermelons that resulted in illness to 998 persons in California (Marshall, 1985). Of serious concern also was (and is) the hazard of pesticide "drift," accidental movement of a toxic compound from the site of application onto susceptible plants, wildlife, domestic livestock, or bees, or into water supplies intended for fisheries or human consumption. Such difficulties prompted scientists, growers, and the public to take a second look at pest management programs that relied exclusively on regular, repeated applications of broad-spectrum chemical poisons. Growing awareness of environmental concerns among the general public provided political support for efforts to incorporate biological, cultural, and physical control techniques more fully into pest management programs. More recently, added impetus for reducing dependence on chemicals has resulted from the increased expense of insecticide manufacture and application. Most chemical insecticides are derived from petroleum products, whose increased base cost (despite the recent decline) is known to nearly everyone. Moreover, developmental costs of insecticides have increased markedly due to greater complexity in manufacture and greater rigor in testing (Chapter 7). Today there exists a need for ever-greater efficiency in agricultural production, and both producers and consumers wish to avoid unjustified use of chemical pesticides.

The dangers inherent in overdependence upon insecticides are particularly evident in developing nations, whose farmers, public health officials, and federal governments usually are struggling together to develop greater agricultural efficiency via increased mechanization, often involving a single widespread cash crop, such as coffee, cotton, or cane sugar, to generate desperately needed foreign exchange. The following five "phases" have been recognized in the development of agricultural pest management in such places (Smith, 1969; Metcalf & Luckmann, 1982).

1. *Subsistence*. This is small-scale, peasant-style agriculture, producing food for humans and livestock, and most crops are consumed locally. Pest control techniques are crude, labor-intensive, and limited in effectiveness, yet a substantial amount of control is realized simply through variation in cropping systems and the inherent tendency for localized "traditional" varieties of crops and livestock to resist attack by insects and pathogens. Subsistence probably has been the prevailing mode of agricultural production for most of human history and is still widely practiced, predominantly in tropical regions but also to a

limited extent in highly industrialized nations, such as those in Europe and North America.

2. *Exploitation.* Usually following the subsistence phase there occurs an enlargement and intensification of area under cultivation, with a concomitant shift to production of cash crops, increased mechanization, and more widespread use of chemical pesticides and fertilizers. During the exploitation phase, the landowner's main concern is to produce profits from the land. The plantation economy of the southern United States was an early instance of exploitation, and farming-for-profit was adopted throughout most of the industrialized nations by the early 20th century. The ready availability of chemicals and farm machinery has converted exploitation into a worldwide trend during the past 40 years.

3. *Crisis.* This results from failure of intensive, mechanized, chemically based agriculture (though the crisis phase sometimes occurs in subsistence agriculture as well). In a crisis, production no longer yields a profit, due to insecticide resistance, generation of secondary pest problems, soil erosion, or increasing salinity. Crisis also results from changes in market conditions, especially falling farm prices coupled with inflated production costs, perhaps with no loss of actual yield. (This may not be due to failure of pest management, but the result is the same: economic failure in agriculture.)

4. *Disaster.* The production system collapses, partially or totally, and the crop can no longer be grown at a profit (if indeed it can be grown at all). Social disruption may follow, as abandonment of land forces ever more rural people into crowded urban areas, intensifying problems brought about by demographic changes during the exploitation phase. The coming of the boll weevil into the southeastern United States is an example of a disaster brought on by a new, unmanageable pest. The collapse of cotton production in northeastern Mexico in the 1960s is another example—in this case, a result of increased salinity and resistance of *Heliothis* spp. to insecticides.

5. *Integrated Management.* After disaster, it may be possible to devise less environmentally disruptive, more balanced management techniques, based on a thorough understanding of pests' biologies and relationships to other organisms and the physical environment of the agricultural ecosystem. (The boll weevil played a major role in stimulating the diversification of agriculture in the southeastern United States.)

The goal of integrated pest management is not to turn the clock back to yesteryear but rather to turn it forward, to make best use of current knowledge and techniques, from efficient computer simulations to more accurate and dependable pesticide application equipment. Practitioners of pest management recognize that increasing efficiency of mechanization in agriculture is a trend that will continue, and that management techniques must be harmonized with the primary motivation of producers to gain profits (except, of course, in the socialist economies). In terms of these five "phases" of agricultural development, pest management seeks to avert crisis and disaster, and to continue realizing a profit while minimizing exploitation.

Reestablishing the balance between insect pests and ourselves with the least disruptive tactics possible is what this book is all about. Insect pest management is but part of a broader field of total crop ecology and management, and the entomologist must constantly keep this broader picture in mind; insect management, if it is to be useful at all, must fit into the entire agricultural production system.

What Is a Pest?

A pest has been variously defined as (1) an animal or plant whose activities interfere with human health, convenience, comfort, or profits (Horn, 1976), (2) a creature that reduces availability, quality, and/ or value of a resource (Flint & van den Bosch, 1981), or (3) an animal or plant whose population density exceeds some unacceptable, arbitrary threshold level, resulting in economic damage (the definition used in this book). Broad definitions of "economic injury level" apply to public health pests and nuisance pests whose population densities exceed an "aesthetic injury level" (Olkowski, 1974); Chapter 2 contains fuller discussion of this. Central to all definitions of "pest" is that the definition is arbitrary, a matter of opinion. Without people, there are no pests. Where there are people, there are pests also, but the decision is often philosophical and personal as well as economic; one man's pest might be another's pet. We may generally agree that *Anopheles albimanus*, a principal vector of malaria in the Neotropics, is a serious pest, but we may disagree on the pest status of an insect whose effects are casual, perhaps merely cosmetic. For example, craneflies (Tipulidae) are attracted to lights inside buildings but do no serious damage, yet many people erroneously consider craneflies to be capable of great harm because of their resemblance to mosquitoes. The decision as to what constitutes a pest may be deep-rooted in a particular value system. In Western Europe and North America there prevails a general anthropocentrism bordering on arrogance that considers all insects as a "lower" form of life that we are justified in loathing and that must be controlled. This popular opinion is often accompanied by an irrational and obsessive fear of creatures with six (or more) legs. (One need only observe the audience's reaction to the first reel of *Indiana Jones and the Temple of Doom* to appreciate the pervasiveness of this attitude. See Mertins, 1986, for additional examples.) Persons who hold such a view to the extreme are understandably reluctant to embrace a philosophy of management that regards pests and their damage as inevitable and considers that the best long-term strategies for their management attempt to keep their numbers in check below damaging levels while recognizing our inability to eradicate pests. White (1967), Perkins (1982), and Frankie and Koehler (1983) discuss more fully attitudes toward pests and their control.

What Is Pest Management?

Definitions of "pest management" are likewise arbitrary and in a state of flux as insect management passes through the transition from nearly exclusive reliance on chemical control to a broader, holistic, multidisciplinary approach (Apple & Smith, 1976; Allen & Bath, 1980). Rabb and Guthrie (1970) defined "pest management" upon its early emergence as a discrete approach to control:

> Pest management is reduction of pest problems by actions selected after the life systems of the pests are understood and the ecologic as well as economic consequences of these actions have been predicted, as accurately as possible, to the best interests of mankind. (pp. 2–3)

A 1977 U.S. Department of Agriculture (USDA) report defined "pest management" as a recognition of the need to balance environmental, economic, and sociopolitical factors in devising control tactics; pest management is

> a desirable approach to the selection, integration, and use of methods on the basis of their anticipated economic, ecological, and sociological consequences. (Allen & Bath, 1980, p. 658)

Flint and van den Bosch (1981) emphasized reliance on natural controls as much as possible; pest management is

> an ecologically-based pest control strategy that relies heavily on natural mortality factors such as natural enemies and weather and seeks out control tactics that disrupt these factors as little as possible. (p. 6)

Perkins (1982) pointed out that current approaches to pest management display three distinguishing features, first articulated by Stern *et al.* (1959).

1. Insect pests, as well as plant pathogens and weeds, are integral components of ecosystems. Interactions among insects, plants, pathogens, and natural enemies must be considered in some depth when devising management systems. Similarly, a management technique, particularly a broad-spectrum pesticide or other traumatic manipulation, such as tillage, burning, or draining, exerts effects whose consequences ramify throughout the ecosystem, sometimes with undesirable (and unpredictable) impact. Risser (1985) summarized this viewpoint: "Agroecosystem management demands an understanding of ecosystem behavior and how natural processes are modified by agricultural objectives."

2. An economic injury level must be established, because not all potential population densities of an insect (or other pest) are likely to cause enough loss to justify management efforts. As will be seen, there is raging debate over how precisely we can define economic injury levels in situations where insect populations, crop yields, farm profits, and costs of production fluctuate widely and sometimes unpredictably. Moreover, an insect known to cause damage in one region might justifiably be regarded as a "potential pest" where it is less abundant, or even absent. The Mediterranean fruit fly ("Medfly") and the Japanese beetle both cause substantial economic loss where they are well established. Many millions of dollars have been spent to eradicate each from California, because of the potential for damage that would exist should they become established. In effect, the California Department of Food and Agriculture regards the economic injury level for both Medfly and Japanese beetle as equal to one, if that one is a fertilized, gravid female, and the long-term benefits of a localized eradication program clearly exceed the cost. (The concept of economic injury level hardly applies in eradication, the goal of which is to eliminate a pest in order to avoid a need to develop a pest management program. Chapter 11 discusses this in more detail.) Assessment of pest population density, ideally by a trained specialist, assists the grower/manager to properly schedule control procedures.

3. Pest management uses selective control techniques, management procedures that pinpoint the vulnerable life stages of a pest, with minimal deleterious impact on beneficial organisms in the associated ecosystem. These techniques may include introduction or augmentation of biological control agents, use of resistant crop varieties, application of narrow-spectrum insecticides, more efficient timing, dosage,

and application of short-residual chemicals, and/or timely planting, harvesting, and tillage.

The key to successful insect pest management is therefore a practical analysis and assessment of an insect's population dynamics in relation to ecological and economic factors impinging upon it. Appropriate and effective management techniques cannot be developed without a conceptual model of the system involved. This depends on a thorough understanding of the pest species' ecology, of key factors in its life cycle, and of potential weak links that might be exploited in a management scheme without producing overly deleterious side effects. The more precisely a modeled description reflects real events, the better it is.

The organization of this book follows this line of thought. Economic injury level is defined and discussed (Chapter 2), and chapters on sampling, modeling single-species populations, and the ecology of insect populations in ecosystems follow. Subsequent chapters develop tactical approaches to insect management: chemical, biological, and cultural manipulations. Finally, some case studies are introduced to illustrate complexities involved in practical pest management.

The ideal of pest management introduced here is one for which scientists strive, yet practical limitations often prevent understanding of a pest's ecology beyond a superficial level. Further research must be done even on our common and familiar pest species. It is also true that some insect populations are so variable as to be thoroughly unpredictable, given our current state of knowledge. Others are sufficiently noxious as to warrant control measures at all times; their economic injury level is *always* below their average population density. Cockroaches infesting areas where human food is processed might fall into this category. Such cases remind us that there remain many instances in which substantial reliance on application of broad-spectrum chemicals is still the only practical approach to insect pest management.

ADMINISTRATION OF APPLIED ENTOMOLOGY

Insect pest management is intimately related to the persons and institutions who make management decisions. In the United States and elsewhere, the roles of government, industry, agricultural experiment stations, universities, and individuals are intertwined into marvelously complex and often misunderstood processes, accompanied by a Byzantine panoply of agencies and acronyms. The discussion below modestly attempts to unravel and demystify part of the process. The bias is toward activities within the United States, and reflects the author's experience, though (for good or ill) the United States serves as a model for administration of such activities throughout much of the world.

Entomological Research

Research is supported by federal and state funds and is conducted by personnel at a network of laboratories within the USDA and the numerous State Agricultural Experiment Stations. Within the federal service, it is the USDA–ARS (U.S. Department of Agriculture, Agricultural Research Service) that staffs and finances laboratories scattered around the nation and at a few locations overseas. For instance, the USDA maintains biological control laboratories in Argentina, France, Italy, and Korea. Most often, the federal laboratories are dedicated to insect problems of regional interest, so that, for example, the USDA–ARS lab at Stoneville, Mississippi, concentrates effort on insect pests of cotton, the most important economic (and political) local crop.

Within the USDA–ARS there are two "program areas" that administer entomological research. Agricultural entomology is overseen by the Plant Science and Natural Resources Program, while insects affecting man and animals are under the Animal and Postharvest Science Program. The U.S. Forest Service (itself a subdivision within the USDA) also maintains and staffs several entomological research laboratories.

At the state level, the Morrill Act of 1862 initiated a system of land-grant universities dedicated initially to instruction in "the agricultural and mechanic arts." Most land-grant universities have evolved into comprehensive research and educational institutions of international stature. The Hatch Act of 1887 provided for the State Agricultural Experiment Stations, which (in most states) have become closely allied with, if not a part of, the state colleges of agriculture within the land-grant universities. The Smith–Lever Act of 1913 established the Cooperative Extension Service (CES), jointly funded by the states and the federal government. The CES is the public education arm of the land-grant university system, and its responsibility is to disseminate research findings from the State Agricultural Experiment Stations to growers and the general public in a form useful to enhancing agricultural productivity and the overall quality of life. Historically, the land-grant universities have developed along traditional disciplinary lines; entomology, plant pathology, horticulture, agronomy, etc., are administered as separate academic departments. It has been suggested that such disciplinary specialization inhibits research oriented toward integrated systems, which has become paramount with the modern emphasis on integrated pest management and ecosystem concepts in agriculture (Altieri *et al.*, 1983).

Funding for pest problems is a joint federal–state venture with the majority of federal funds disbursed according to so-called "formula funding." Formula funding is a rather complex arrangement whereby the number of farms (rather than total population or total farm productivity) is emphasized in determining the amount of funding received by each experiment station from the federal government. For instance, Ohio and Mississippi have a large number of (small) farms, and receive a proportionately large share of federal agricultural research funds. Formula funding has been criticized as presumably lacking sufficient stimulus for incentive in research. In an effort to counter charges that agricultural research is relatively unproductive when supported by federal formula funds, the USDA recently initiated a program of competitive grants (administered by the Competitive Research Grants Office, CRGO), available to research scientists from all U.S. institutions (including the USDA's own research laboratories). About two thirds of CRGO funding continues to be awarded to state colleges of agriculture and agricultural experiment stations, as one might expect by virtue of these institutions' mission to investigate problems of importance to agriculture.

Overall research funding is determined by needs that are evaluated both by administrators and by scientists in the field. The USDA also has a role in coordinating research at the state level through the Cooperative States Research Service (CSRS), which funds regional research projects simultaneously involving scientists at several State Agricultural Experiment Stations. For example, in the North Central region, there is (among others) a regional project on biology and control of corn rootworms, a pest of considerable impact on corn production throughout the Midwestern area. Regional research committees function as technical committees to suggest potential areas needing work, and these programs are endorsed by administrators (department chairmen, directors of the State Agricultural Experiment Stations, etc.). The suggestions are incorporated into annual reports and budget requests by state and federal personnel. For instance, the relatively sudden failure of some soil insecticides to effectively control corn rootworms in conservation tillage systems in the midwestern United States is a problem with potentially serious implications for a major regional agricultural commodity. Scientists and administrators have suggested to the Congress and President (via the USDA) that more resources be diverted to research on soil ecology and microbiology in this region.

Inevitably, old-fashioned politics also partially determine research and development priorities in applied entomology. (This is not fundamentally different from any other human endeavor on which public funds may be spent.) The gypsy moth is an example. In the northeastern United States the gypsy moth defoliates hardwood trees, including those surrounding and shading the dwelling places of wealthy and powerful persons. Damage is highly visible, and it is therefore rather simple to generate political enthusiasm (and controversy) during outbreaks. The USDA has responded to pressure by initiating major research programs on the gypsy moth. Of course, interest runs highest during outbreak years and tends to dissipate just as does the moth itself during intervening years. Due to the lag in appropriations, there is a time delay in the funding cycle not unlike coupled predator–prey oscillations, to be discussed in Chapter 5.

A substantial proportion of entomological research is undertaken by industrial concerns. This work encompasses not only direct product development but research on fundamentals of insect physiology and biochemistry in an effort to understand insect function and to search for ecological and physiological "weak links" in insect biology that might be exploited more effectively for pest management. Foremost in effort and resources directed toward these endeavors are the huge multinational agricultural chemical corporations, each of which manages at least one major research facility within the United States and often has overseas branches as well. Financial return is the primary goal of these corporations, a fact that seems not always evident to critics of industrial research and development. The early arguments of Carson (1962) against pesticides and the agricultural chemicals industry have been modernized and expanded by several other critics (e.g., van den Bosch, 1978; Dethier, 1976). It has been variously stated that there exists a "pesticide conspiracy" wherein agribusiness, multinational corporations, the USDA, and state universities and experiment stations are somehow in league with one another to perpetuate the use of agricultural chemicals while quashing the aspirations of those who hope to develop alternative pest controls. The debate becomes emotional, and sometimes the rhetoric of class struggle and Marxist philosophy is interjected, with the expected polarization of viewpoint and sloganeering. There is no doubt that the economic system prevalent in the United States and most of the nonsocialist world encourages the development and use of the broad-spectrum chemical solutions to insect pest problems. (Economic pressures and ecological shortsightedness also encourage widespread use of broad-spectrum insecticides in the Soviet Union and its satellites.) Whether there is open or covert conspiracy is certainly not clear, though the matter can be explained via simple economics. A true "pesticide conspiracy" might well take a good deal more coordination and finesse than seems to prevail at present among the alleged conspirators.

Regulation

Enforcement of pesticide laws, including registration of pesticide labels and certification of pesticide applicators, is jointly the province of federal and state authorities—specifically, the U.S. Environmental Protection Agency (USEPA) and state departments of agriculture (or, in a few cases, state EPAs). Most states have pesticide applicator laws more or less consonant with federal law.

State agencies also cooperate with federal authorities in insect (and weed and pathogen) surveys and enforcement of quarantines (Chapter 11). These activities are coordinated by the USDA–APHIS (Animal and Plant Health Inspection Service), operating under authority of the Plant Quarantine Act (1912) and the Federal Plant Pest Act (1957). APHIS is an arm of the marketing division of the USDA and *not* of the USDA–ARS, a bureaucratic fact that may confuse the uninitiated. APHIS programs are not intended to be research programs per se, but rather assist in pest management decision-making by monitoring the numbers and movements of significant pests.

Overseas Entomological Activity

Most economically developed nations of the world have national or federal research and educational institutions similar to the USDA–ARS and the CES. Examples are the Commonwealth Scientific and Industrial Research Organization (CSIRO) in Australia, and the Commonwealth Institute of Biological Control (CIBC), in nations of the British Commonwealth. There are also several major international laboratories that concentrate on regional research problems. Examples are the International Rice Research Institute (IRRI) in the Philippines, the Centro Agronómico Tropical de Investigación y Ensenanza (Tropical Agricultural Research and Training Center, CATIE) in Costa Rica, and the International Center for Insect Physiology and Ecology (ICIPE) in Kenya. The United Nations supports international research and educational activities through the Food and Agricultural Organization (FAO, headquartered in Rome, Italy) and World Health Organization (WHO, headquartered in Geneva, Switzerland). These organizations are committed to increased effort on developing local talent and local solutions for local problems. Without extensive modification, the energy-intensive and mechanically intensive agriculture typical of the United States and Western Europe is inappropriate for much of the world.

PROFESSIONAL SOCIETIES

Entomologists, sharing a common interest, have found it helpful to band together into professional societies. In the United States the largest and most influential is the Entomological Society of America (ESA), headquartered in College Park, Maryland. The ESA has about 9,000 members, representing academic institutions, state experiment stations, governmental agencies, and industry, as well as students and dedicated amateurs. It was formed in 1953 from a merger of the "old" ESA (primarily those who were interested in basic entomological research) and the former American Association of Economic Entomologists (AAEE). The merger of two groups with (sometimes) disparate interests has been stimulating and fruitful, though not without challenges. Generally each group has been strengthened by association with the other. The ESA is divided into five geographic branches (Eastern, Southeastern, North Central, Southwestern, and Pacific), and six subdisciplinary sections. The ESA holds an annual conference (the site rotates among the branches), and each branch also holds an annual meeting. These gatherings are a melange of committee meetings, presentations of the latest research findings by scientists, desperate searches for professional employment by students, and the customary socializing. The ESA operates a placement service, and publishes three professional journals (*Annals of the Entomological Society of America, Environmental Entomology,* and the *Journal of Economic Entomology*), as well as its bulletin, newsletter, and miscellaneous publications. The American Registry of Professional Entomologists (ARPE) is partially supported by the ESA, and meets at the same time and place. It exists for those who desire the benefits of professional registry that may enhance their employment.

In addition to the ESA, many states have regional and local entomological societies, though their role in pest management decision-making is generally secondary to the more academic aspects of professional entomology. An exception is the Association of Applied Insect Ecologists (AAIE), which is based in California and whose functions are similar to those of ARPE.

Other nations generally have at least one national entomological society, often supported by the government. (ESA is not.) At 4-year intervals an International Congress of Entomology is held; the XVIII Congress is slated for Vancouver, British Columbia, in 1988.

2

INSECT PESTS AND
ECONOMIC DECISIONS

A critical first step in pest management is the decision as to whether or not an insect is indeed a pest. During the past decade, considerable effort has been invested in defining economic injury levels and attempting to determine why one species becomes a pest while another does not. Determination of economic injury levels remains in flux as entomologists and agricultural economists struggle, cooperate, and argue about precision, uncertainty, definitions, population ecology, agricultural practices, market analysis, and impact of insects on yields. This chapter develops some of the recent ideas about what constitutes a pest, and amplifies on the simplistic notions of pests introduced in Chapter 1.

ECONOMIC INJURY LEVEL

Simplified View

In Chapter 1, a pest was defined as any animal or plant whose population density exceeded some arbitrarily defined threshold beyond which it interfered with human health, comfort, convenience, or profits. This was formalized graphically by Stern *et al.* (1959), as illustrated in Figure 2-1, and represents a concept of economic injury level held by most agricultural entomologists (Mumfort & Norton, 1984). Figure 2-1A depicts a traditional, familiar instructional model of population growth and regulation (developed more fully in Chapter 4). Upon establishment in a favorable environment, any population increases in

(A)

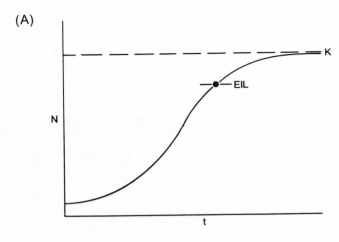

(B)

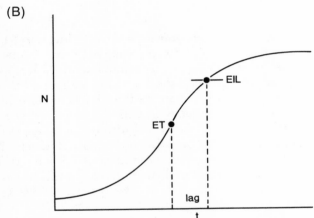

FIGURE 2-1

Economic injury level. (A) Sigmoid population growth, describing the relationship between insect density (*N*) and time (*t*) for a population newly introduced into a favorable environment. EIL = economic injury level, above which insect is considered a pest; *K* = environmental carrying capacity, the maximum value around which population density fluctuates (Chapter 4). (B) Same relationship as in A. ET = economic threshold, density at which control measures are applied to prevent *N* from exceeding EIL.

density to an upper limit (K in the figure), which it does not exceed, due to predation, competition, or other environmental factors. K is called the "carrying capacity," or "environmental resistance," in classical ecological studies. (As will be seen, reality is much more complicated than this model suggests.) The economic injury level (EIL throughout this book) is an arbitrary density lower than K; at densities higher than the EIL the insect (or other organism) is considered to be a pest. EIL has become a nearly universally accepted term, though some (e.g., Pedigo *et al.*, 1986) advise that "injury" properly refers to abnormal growth or development, whereas "damage" refers to loss of value; therefore, the economic "injury" level is really a "damage" level. The rationale of pest management is to alter a system with as little environmental, economic, or social disruption as possible, so that N is consistently less than the EIL. This usually involves an environmental manipulation that may include input such as application of insecticide.

Because population growth is time-dependent, there is often delay after treatment before population density actually declines. Figure 2-1B shows that a treatment, such as application of a microbial insecticide, might be delayed in its impact and therefore must be applied at a point preceding the EIL. This is the "economic threshold" (ET in Figure 2-1B). Applying a treatment at the economic threshold allows time for delayed action, and the pest population continues to increase during the interval between the economic threshold and the EIL. When a fast-acting chemical insecticide is used, the economic threshold may be nearly identical to the EIL. Some publications reflect this by making no distinction between the two, and use "economic threshold" and "economic injury level" interchangeably. This is reasonable when considering insecticides that kill within a day (or less) of application. For growers and pest management consultants it is more critical to know the economic threshold, for it is the population density at which practical action is required. Poston *et al.* (1983) defined the economic threshold as a decision level minimizing the chance that the EIL will be exceeded.

In practice, the relationship between the EIL and pest density is complicated by changes in each. Population densities of insects (or of anything else) generally do not remain constant, but fluctuate in response to intrinsic and extrinsic factors. Sometimes these fluctuations are periodic, regular, and predictable, though at other times they seem chaotic and unpredictable even after exhaustive study. Figure 2-2A depicts an intermediate situation, a population curve representing what might occur during a single growing season when a crop pest has several generations. (This curve is similar to some of my data for

(A)

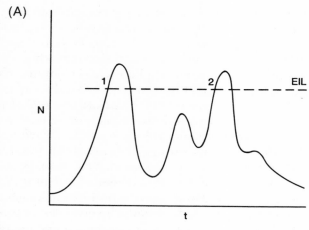

(B)

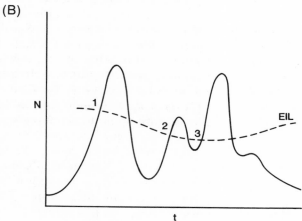

FIGURE 2-2

Growth curve for population of varying density. (A) Constant EIL; pest outbreaks occur at points 1 and 2. (B) As in A, but with varying EIL; outbreaks now occur at 1, 2, and 3.

aphids infesting collards in central Ohio.) The insect becomes a pest only at points 1 and 2; management measures are warranted to prevent further increase at these two times only. Obviously, an ability to predict the arrival of time 1 and time 2 is essential to effective management in this situation.

EILs are not constant, but vary with market conditions, agronomic practices (variety, fertilizer, irrigation, etc.), geographic location, and time. Costs of control and consumer preferences are factors in the determination of the EIL, which invariably has a major subjective component (Mumford & Norton, 1984). Figure 2-2B illustrates this; here, in comparison with Figure 2-2A, there are three pest "outbreaks," and the third (3) occurs slightly earlier than the second outbreak of Figure 2-2A, wherein the EIL was constant.

The simplistic models depicted here are deterministic (Chapter 4); that is, they assume single, predictable values for what, in reality, are variable and uncertain functions. For example, the EIL may change in response to plant stress (Chapter 6). A reflection of this uncertainty is that relatively few EILs have been determined and published (Poston *et al.*, 1983). Insect population densities and economic conditions might be described more legitimately by distributions rather than by single values. Figure 2-3 illustrates this to an extreme, yet growers and entomologists alike must give some consideration to the full range of potential conditions, as well as to averages. In highly unpredictable situations the determination of the EIL may be sufficiently uncertain as to be of little practical use. This is especially likely in urban ecosystems, wherein measurable economic factors are less important than are subjective levels of people's tolerance for insects and evidence of their damage (Frankie & Koehler, 1983). The so-called "aesthetic injury level" (Olkowski, 1974) may vary over an extremely wide range (Chapter 12). This suggests an advantage to reducing the interval

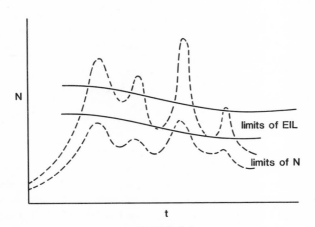

FIGURE 2-3

Variable population density and economic injury level. Both population density (*N*) and EIL fluctuate between limits depicted, adding uncertainty and greatly complicating decision on when to take action against pest.

between economic threshold and EIL in order to alleviate uncertainty, and a fast-acting chemical insecticide is most likely to do this (Pedigo *et al.*, 1986).

The concept of EIL developed here has marginal utility to pests of importance to public health, wherein the most critical factor is the presence, or even the potential, of a pathogen within the vector or host population. The population density at which a pathogen is maintained by the vector can be considered to be the EIL. An economic threshold should be established well below this, for humanitarian reasons as well as because of the potential for societal disruption due to outbreak of a major disease epidemic. (For hard-core realists who insist upon economic analyses, Blomquist, 1979, estimated the value of an American human life in 1978 to be $370,000; Pimentel *et al.*, 1980, estimated the value to be $1 million; and each figure is probably too low.)

Marginal Analysis

The "marginal analysis" approach first proposed by Hillebrandt (1960) and refined by Headley (1972, 1975) is here introduced as an illustration of an agricultural economist's view of EILs (Figure 2-4). This analysis develops the EIL as an optimum value from an incremental function. Figure 2-4A depicts the relationship between the value of a crop and the density of a potential pest. The value, in dollars (or pounds, francs, rubles, etc.), is assumed to be constant and at a maximum between pest densities zero and X. If pest density increases beyond X, the value of the crop declines at an increasing rate, through point Y, at which the crop is worthless ($\$ = 0$). Conceivably, the pest density might increase beyond Y, because the crop, though worthless economically, might support even more insects. The lower portion of Figure 2-4A shows the increasing rate of change in value (in the terminology of calculus, the lower curve is the derivative of the upper curve).

Figure 2-4B shows the relationship between cost of control per pest (C) and insect density (N) for any desired level of control. To achieve extremely low densities of a pest is very expensive. For example, the state of California spent several million dollars to eliminate perhaps a few thousand Mediterranean fruit flies in the early 1980s. High densities of pests, on the other hand, can be achieved at low cost simply by doing nothing when they appear. In the extreme, I can ensure the low-cost occurrence of thousands of Mexican bean beetles in a central-Ohio vegetable garden by investing 79 cents in a packet of

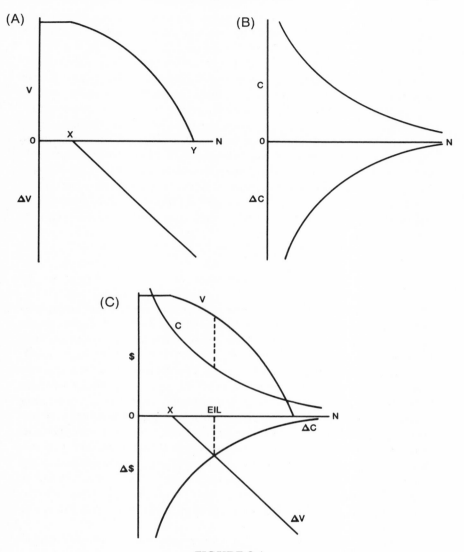

FIGURE 2-4

Incremental economic injury level. (A) Relationship between crop value (V) and pest density (N). At $N = X$, value begins to decline; beyond $N = Y$, the crop is valueless. (B) Relationship between cost of control (C) and pest density (N). (C) Value and cost combined; EIL = economic injury level, point at which difference between V and C is maximized (see text for fuller explanation). From Headley (1975). © John Wiley & Sons, Inc. Used with permission.

bean seeds (and planting them) but making no subsequent effort to control the inevitable beetles. The lower portion of Figure 2-4B is the rate of change in cost with increasing density, analogous to Figure 2-4A.

Figure 2-4C combines value (2-4A) and cost (2-4B) into a single relationship between cash and pest density. According to this model, the difference between value realized and cost of control is maximized precisely where the incremental change in value equals the incremental cost of control—that is, where the two lower curves intersect. At this point, a grower's return on cost is maximized. In a perfectly functioning capitalistic economic system, management measures should therefore be taken at this point, and therefore this is the (idealized) EIL. Note that "injury"—that is, evidence of pest activity—may be visible at point X but there would be far less return on investment should control measures be applied at point X, due to the high cost of control.

Few actual measurements of crop values and management costs in relation to insect densities have been made in the field, though some studies have yielded approximations that are likely to be more useful in decision-making than is the guesswork that formerly prevailed. Olfert *et al.* (1985) presented an example of such an approach, estimating impacts of the wheat-stem midge over a variety of densities in the Canadian prairies. Chapter 12 discusses additional examples of EILs in practice.

Figure 2-5 extends Headley's incremental model to reflect changes in farm prices (Norgaard, 1976). A price increase shifts the value curve upward and steepens its slope. This reduces the EIL. On the other hand, an increase in the cost of control, without a concomitant increase in crop value, raises the cost curve, but, if the incremental cost of control does not change, there is no change in the EIL. The farmer's profit margin is simply reduced by increased cost of control.

Additional Considerations

There are other models for EIL, but Stern's and Headley's have gained widest acceptance. All models for EIL consider market value of crop, cost of control, amount of injury caused per pest, and the crop response to pest populations of various densities (Poston *et al.*, 1983). As one might expect, the difficulties encountered in determining EILs are compounded for pest complexes in which more than one pest species is simultaneously active. This seems more often to be the rule rather than the exception in agriculture. The problems are compounded in instances in which an agricultural crop generates a pest of medical

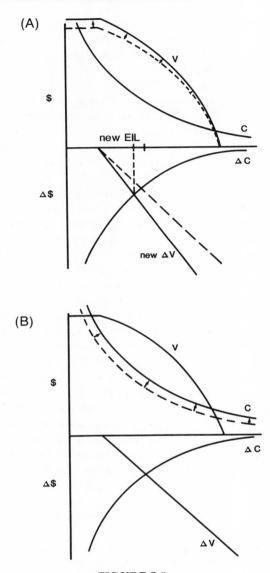

FIGURE 2-5

Incremental EIL, showing effects of (A) increase in farm prices (the change in value, ΔV, is steepened, lowering EIL) and (B) increased cost of control (EIL remains unchanged, but profit margin—difference between C and V—is reduced). From Headley (1975). © John Wiley & Sons, Inc. Used with permission.

importance, such as the emergence of mosquitoes in large numbers from rice fields (Lichtenberg & Getz, 1985).

Most models for EIL assume deterministic values for crop value and cost of control, whereas the EIL is often a probabilistic (stochastic) function. In capitalistic economies, producers of agricultural products are in business, and to remain solvent they must deal with the tradeoffs involved in simultaneously realizing adequate profit while avoiding excessive risk. The farmer's objectives may include reducing variance in profit, as well as maximizing profit. Devastating floods, unseasonable freezes, or outbreaks of blight and locusts may occur with very low frequency but these events do occur, and, when they occur, the prudent businessperson who had planned to survive such extremes is more likely to remain in operation than would someone who had not planned for the unusual. Such considerations may tempt growers away from treating only when the EIL is exceeded and toward use of preventive "insurance" treatments, often with a broad-spectrum insecticide, to kill as many different kinds of actual and potential pests as possible (Feder, 1979). (This philosophy is often reflected in recommendations from the CES, whose specialists may recommend insecticide treatments because they would rather err on the side of caution, despite extra expense to the grower.) This effectively increases the cost of control and reduces profit margin while increasing the probability that economic injury will not occur at all. Norton and Conway (1977) pointed out that the avoidance of risks is especially critical both for farmers of very small holdings and for large corporate farms. The former have few alternatives in the event of crop failure, whereas corporate farmers have too great an investment, and neither wish to take chances that might be acceptable to the manager of a medium-sized, diversified farming operation. On the other hand, mid-sized farms (annual net income around \$10,000 to \$20,000, in the United States) tend to have the most severe cash-flow problems (Tweeten, 1983). The part-time or avocational farmer or gardener has the greatest range of options, and in the event of total crop failure he or she might simply head for the nearest grocery store, an option usually unavailable to one who farms for a livelihood.

One means of dealing with the uncertainties inherent in pest management is via a "decision-tree" model of EIL, simultaneously depicting maximum profits and the probability of avoiding risk. Figure 2-6 shows data for the gypsy moth as an example (Valentine et al., 1976). Here, the manager of a forest is assumed to have three options: C_1, do nothing, at no cost, C_2, bacterial insecticide at \$20/hectare, or C_3, chemical insecticide at \$27/hectare (including both cost of direct application and impact on the ecosystem). There are three potential

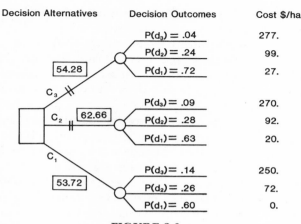

Decision Alternatives   Decision Outcomes   Cost $/ha

FIGURE 2-6

Decision tree for gypsy-moth management. C_i = control options: 1, do nothing; 2, microbial insecticide; 3, chemical insecticide. $P(d_i)$ = probabilities of various defoliation levels: 1, light; 2, moderate; 3, heavy. Figures in boxes are average expected cost of each option (see text for fuller explanation). From Valentine *et al.* (1976). © Entomological Society of America. Used with permission.

levels of defoliation: light (d_1), with no loss; moderate (d_2), with an average loss equal to $72/hectare; and heavy (d_3) with an average loss of $250/hectare. Through exhaustive study, the probabilities (P_i) for the three levels of defoliation are determined for each treatment, and are entered in the decision tree along with the prospective costs (simply equal to the treatment cost plus the loss due to defoliation, d_i). Each cost is multiplied by its probability and results are added within each C_i for an estimate of the average cost of the treatment. In this example, bacterial insecticide became the most expensive option ($62.66), whereas the cost of chemical insecticide was approximately equal to that of doing nothing. The most critical difficulty in applying decision-tree models to operational pest management systems is accurate determination of the probabilities, for an apparently minor change in probability might result in an enormous increase in average cost. Plant (1986) discusses additional examples of the relationship of uncertainty to EIL.

"Classical" economic analyses (e.g., Headley's) normally do not consider the effects of so-called "social costs" of pest management decisions. Such problems as acute and chronic poisoning, costs of legal regulations, and evolution of insecticide resistance are all considered "externalities" by classical economic analysis. If these were to be figured into the growers' balance sheets (perhaps by requiring payment into a fund to indemnify victims of accidental poisoning or to clean up after spillage endangering fish and wildlife), the cost of control would simply increase (in Headley's model), moving the cost curve (Figure 2-5B) upward and reducing the grower's margin of profit.

Some have suggested that widespread adoption of integrated pest management might increase agricultural yields. Headley (1975) pointed out that increased yields achieved via improvements in pest management (or for any other reason) exacerbate the troublesome paradox of North American agriculture: higher yields add to surplus production and further depress farm prices. (In 1985, a production increase of 1% resulted in an approximately 4% decline in prices.) One commercial goal of pest management is to increase the efficiency of production, ideally increasing the profit margin by reducing the cost of management (Figure 2-5B). (In some cases, adoption of pest management has led to an increase in insecticide application. Control of the potato leafhopper on alfalfa in the midwestern United States is an example. See Chapter 12.)

To address adequately the role of management techniques in determining the EIL, an interdisciplinary market analysis should be undertaken to determine the fiscal and social costs of various management tactics. Such analyses are in their infancy; however, in the relatively few studies published to date, it seems that unless economic damage *always* occurs in the absence of insecticide, there is an advantage to investment in a pest management program wherein fields are "scouted" and insecticides are used only on an "as-needed" basis—that is, when $N >$ EIL. For example, White (1979) found that apple producers in New York saved \$12 to \$16/acre when paying a scouting fee of \$12/acre (Table 2-1). Greene *et al.* (1985) estimated an average savings of \$9/acre to soybean growers whose crops were monitored for pests versus those who simply applied insecticide according to a predetermined schedule. Monitoring is more efficient (in terms of cost per hectare) when scouting involves a small number of larger fields rather than a larger number of smaller fields. This is yet another economy of scale providing additional momentum to the nearly universal trend toward larger-scale and more mechanized agriculture in industrialized nations.

Confounding such analyses as the preceding is the great variation in both insect populations and growers' acceptance of management programs. Burrows *et al.* (1982) found a trend toward reduction in costs of insecticides when cotton producers received information on

TABLE 2-1

Effect of Pest Management Monitoring Program on Costs of Management Decisions in New York Apple Production[a]

| | Pesticide cost/acre | | |
	Conventional preventive spray	Pest management scouting	Savings/acre[b]
Fresh fruit	$90	$66	$12
Processing	$90	$62	$16

[a]White (1979).
[b]Profit after scouting fee.

insect management from agricultural consultants rather than from salespeople for insecticide companies, though variation in responses masked any definitive, statistically significant differences. Gruys (1982) found that resistance of growers to insect monitoring programs was the chief impediment to acceptance of these programs for apples in the Netherlands despite an average savings of Dfl 13/hectare. Mumford and Norton (1984) speculated that many growers simply did not wish to make frequent management decisions but would rather adopt scheduled, standard treatment procedures. In England, Mumford (1981) found that, despite a well-researched and widely publicized EIL of four green peach aphids/plant (British Sugar Corporation, Ltd.), a randomly selected sample of 60 farmers used 23 different EILs. In most cases, insecticides were applied as a precaution, well in advance of damaging numbers of aphids. Aesthetic EILs are particularly subject to such variation, as noted earlier. For several years I have asked undergraduate and graduate students, mostly entomology majors, about their personal levels of tolerance to German cockroaches in a classroom at The Ohio State University. Table 2-2 shows their responses to a logarithmic progression of suggested EILs. Students of entomology might be expected to exhibit a higher tolerance to insects than would the general public. (Subsequent to questioning, the students discover that the classroom is indeed cockroach-infested, whereupon the EIL usually drops to $N = 1$, and any cockroach that appears stands a good chance of being forcefully compressed between two solid surfaces—foot and floor or notebook and bench top. Zungoli and Robinson, 1984, found that public tolerance of German cockroaches was generally high initially but declined once management procedures were instituted.)

TABLE 2-2

Proportion of Students Enrolled in Insect Pest Management Course Tolerating German Cockroaches at Various Densities in Classroom, The Ohio State University, 1972–1985 ($n = 125$)

Cockroach density	Proportion of students with EIL exceeded
0	0
1	0
2	<1%
4	10%
8	25%
16	75%
32	95%
64 or more	100%

About 33% of all pesticides used in the United States are applied in urban environments where costs of management are highly concentrated and visible, especially in community-wide pest control programs. Benefits, on the other hand, may be more diffuse, and less visible, and decision-makers may simply use insecticides as the easiest alternative available (LeVeen & Willey, 1983).

EXPANSION OF TIME FRAME

All management decisions are based on uncertain and incomplete information, and this uncertainty increases as time delays are lengthened. For instance, livestock may take 2 or 3 years (or longer) from birth to reach market, and a producer of beef cattle might well ponder the impact of a fly-induced weight loss today on a beef market 2 years from now. On a much longer time scale, red oak borer larvae bore tunnels in the trunks of young trees 10–20 centimeters in diameter. Tunneling in stands of young trees does not affect quality of lumber until the trees are harvested, which may be up to 40 years later (Donley, 1976). A producer of lumber may face a choice between investing in high-yielding bonds and mutual funds or buying the insecticide necessary to improve the quality of timber that may be valuable 40 years in the future. The first option may be more prudent, given the uncertainty of our knowledge of the demand for oak lumber in the year 2027.

Time delays between input, decision, and application of management tactics are a persistent problem for centrally planned agricultural management typical of some socialist nations. Decisions with respect to planting, pesticide treatment, and so forth are made at a central level with substantial "bureaucratic delay," resulting in considerable inefficiency. Of course, this difficulty is not limited to pest management decisions, but it suggests that, in order to be most effective, pest management decisions are best left to the individual grower, farm manager, or homeowner using the best information that can be supplied by professionals.

ROLE OF ENTOMOLOGICAL RESEARCH

The practical determination of EIL is difficult from both entomological and economic standpoints. Ideally, the entomologist's role is to investigate the biology of a pest in sufficient detail so as to determine the probability with which a given set of environmental conditions may lead to increase in the pest's population to, and beyond, damaging levels, as defined by economists. One goal of integrated pest management is to replace the use of unnecessary insecticide with information. At least the entomologist should provide enough information to assist growers to make intelligent decisions and to identify risks, both short-term consequences of not treating and longer-term consequences, such as the evolution of insecticide resistance. This is difficult because year-to-year variability may result in increased pest impact in certain years despite identical pest densities. Moreover, problems of scale may give rise to misleading results. Estimates of insect-caused loss may be greater in smaller test plots than in larger fields that resemble more closely those in actual production (Lamb & Turnock, 1982). Artificial defoliation may not duplicate the actual impact of plant-feeding insects (Chapter 6). It is nearly impossible adequately to estimate potential yield losses when an entire region is totally infested, because no localities devoid of the pest are available for legitimate comparison. For instance, the pink bollworm is found throughout cotton-producing areas of the arid southwestern United States, where there are no uninfested, irrigated fields available for comparative studies (Burrows et al., 1982). It is also a major challenge to differentiate among the confounding effects of weather, pathogens, nutrition, and so on when attempting to estimate losses due to insects or weeds alone. Harcourt (1970) advocated a life table approach (see Chapter 4) to determine EILs. Intensive sampling of crop plants throughout the growing season at various levels of insect infestation could provide

much insight into the specific impact of various pest densities. Perhaps unfortunately, Harcourt's approach has not been widely applied.

Another difficulty lies in assessing the relative harmful and beneficial impacts of some insects that cannot be neatly considered "pests."
Although most persons might find it difficult to argue the positive
value of malaria-bearing anopheline mosquitoes, or corn blight, or
brown rats, the net negative effects of many other pests are not so clear-
cut. For example, *Lygus* bugs at high densities are considered to be
pests by cotton producers because the bugs damage developing squares
and bolls, but the same insects at low densities feed mostly on sap from
lateral shoots, and the resultant thinning can help to produce a high-
yielding crop (Gutierrez *et al.*, 1977a). A low *Lygus* population may
thus reduce a grower's investment in chemical thinning. Imported fire
ants are well-known in the southeastern United States for their painful
attacks on humans and livestock; sometimes they are lethal, especially
to nestling birds, including quail and poultry (Komarek & Kloft,
1980). Moreover, the earthen mounds built by fire ants cause mechanical damage to mowing and harvesting equipment. Yet fire ants are
significant predators of immature stages of the lone star tick (a serious
pest of cattle), and also eat sugarcane borers and boll weevils (Sterling,
1978), among other insects. Insecticide treatments against fire ants in
pasture have resulted in increased densities of horn flies, apparently
after reducing predation by other insects (Howard & Oliver, 1978) in
addition to the fire ants themselves (Summerlin *et al.*, 1984).

In another instance, yellow jackets (*Vespula* spp.) are a common
and increasing urban nuisance in the northeastern and Pacific United
States, and their stings are sometimes sufficiently severe as to warrant
expensive medical treatment (or funerals). Yet, during summer, these
same yellow jackets are voracious predators on caterpillars that would
otherwise damage home gardens and ornamental plants. Though my
observations lack adequate control, I have observed declines in cabbageworms, loopers, cutworms, and tomato hornworms as yellow
jackets increasingly have nested in and around manmade structures in
Columbus, Ohio.

HOW INSECTS BECOME PESTS

As noted earlier, insects (or weeds, pathogens, or vertebrates) become
pests either by their population density's increasing beyond an arbitrarily determined EIL, or by downward adjustment of the EIL in
response to changing attitudes and/or market conditions. Increase in
density can be caused by one or several of many intrinsic and extrinsic

factors (Chapters 4, 5, and 6), ranging from relaxation of the action of mortality factors to change in the insect's biology resulting in local or widespread population increases. In most instances, origins of pest problems can be traced to one or more of the following general causes.

Entry into Previously Uninhabited Regions

Perhaps 40% of the most damaging agricultural pests in North America originated elsewhere. The gypsy moth, Japanese beetle, codling moth, European corn borer, imported cabbageworm, even the "American" cockroach, have all been imported, as have many of our crop plants, mostly from western Europe (Wilson & Graham, 1983). Many noxious weeds and plant pathogens are likewise aliens, along with such vertebrate pests as starlings, house sparrows (both intentionally introduced), and brown rats. Some of these immigrants are considered pests in their native homeland, though often their populations increased greatly when introduced into a new environment lacking species-specific predators and parasites. This premise is the basis for the traditional approach to biological control (Chapter 9), wherein specific natural enemies are selected for colonization in the pest's new homeland. The extensive and sophisticated quarantines administered by various governmental agencies are designed to minimize accidental introduction of potential pests (Chapter 11).

Shifts in Host-Plant Preference Followed by Extension of Geographic Range

Agricultural development may bring a crop variety into close geographic proximity with an insect species that feeds on a closely related plant. Small populations of the potential pest may become established on this new host, which is unlikely to display any physical or chemical resistance. Ultimately the pest comes to prefer the new host, and populations increase to damaging levels. Bush (1969, 1975) concluded that the original hosts of the apple maggot, a North American native, were various species of hawthorn. When apples were introduced during colonial times, populations of apple maggots apparently switched from hawthorn to apple as a preferred host. Similarly, the alfalfa butterfly was originally restricted to the arid Great Basin of the United States, where it fed on wild legumes. When alfalfa was first grown there, the insect expanded its food range to include this new plant (Tabashnik, 1983), and subsequently spread from Atlantic to Pacific. It becomes a pest only in warmer climates with an appropriately long growing season. Yoon and Richardson (1978) suggested that many insect populations evolve with speed sufficient to produce significant genetic changes within a few thousand years. This could lead to evolution of "new" pests in the unlikely but possible event of eradication of a pest over a wide geographical range. The existence of genetically distinct races or "biotypes" in such insects as the screwworm fly (Chapter 11) and the greenbug lends credibility to this speculation.

Change in the Biology of a Host

In breeding crop plants there are compelling economic reasons to produce high-yielding varieties at a sacrifice of other positive traits. In the United States, sweet corn grown during the 1930s had limited tolerance to damage by the corn earworm. The advent of synthetic contact insecticides provided an efficient alternative for controlling the earworm, and at much lower densities than previously attainable. Modern high-yielding varieties of sweet corn are much more susceptible to earworm damage, a matter of recent concern because the earworm has developed resistance to several chemical insecticides. Loss of plant resistance is potentially a very serious concern for the so-called "green revolution," wherein high-yielding but highly susceptible varieties of staple grains (rice, wheat, maize) have replaced hundreds of local, traditional varieties that were (and are) more resistant to insects and pathogens. The newer varieties require heavy inputs of pesticides, inviting the prospect of insecticide resistance. This could lead to the crisis and disaster scenario outlined in Chapter 1.

Changes in Management Technology

The introductory chapter pointed out an increasing trend toward mechanization in agriculture. One result has been a tendency toward monoculture: ever-larger plantings of a single crop species over a wide region. This activity concentrates food plants for crop pests, and some evidence indicates that reduction in habitat diversity reduces numbers of predators at least in some agroecosystems (Altieri et al., 1978; Bach, 1980; Horn, 1981). Widespread use of synthetic organic insecticides has resulted in outbreaks of secondary pests, such as bollworms on cotton and European red mites in apple orchards. Their population densities increased after widespread insecticide treatments were targeted for other pests in the agroecosystem. Indiscriminate insecticide use against

Liriomyza leaf miners, confounded by taxonomic uncertainty and a lack of basic research, has greatly complicated the management of these pests on greenhouse and outdoor vegetables (Parrella & Keil, 1984). Weeds may be a more severe problem in agricultural systems that rely on mechanical harvesting equipment easily jammed by weeds.

Changes in Consumer Preference

Along with the trend toward mechanization in agriculture and (since 1945) increased reliance on chemical insecticides, there has been (in the United States) a steady decline in federal tolerance limits for insect fragments in food intended for human consumption (Pimentel *et al.*, 1977). Such tiny arthropods as thrips, aphids, and mites, though doing little damage in low numbers, are classified as "filth," the same status assigned to rat droppings. Farm products containing insects or their fragments in excess of federally specified levels may be rejected for human consumption. This is a matter of particular concern in the rapidly expanding food-processing industries, whose standards of cosmetic appearance are exceedingly strict throughout the industrialized nations. Particularly in North America and western Europe, consumers have come to expect insect-free, unblemished fruits and vegetables, and are dissatisfied with anything less. Considerable (and, some would suggest, misguided) control effort is directed against insects and plant pathogens to keep them at artificially low levels in order to satisfy the demands of consumers. This extends to ornamental plants also, not only in the wholesale nursery industry (where there are compelling reasons to control pests at low levels lest they be accidentally dispersed along with stock) but also among homeowners. For example, Dowell *et al.* (1979) found that the citrus blackfly at low densities did no direct harm to ornamental trees but deposited sooty mold that, according to a poll of homeowners, "made trees look bad" and warranted control measures. Recall that Zungoli and Robinson (1984) found that tolerance of German cockroaches declined as efficiency of management improved. Such shifts in consumer standards contribute to the estimation that crop losses to insects have remained unchanged (at about 35%) from 1945 until at least 1980 (Pimentel, 1981), despite the widespread use of highly effective and (until recently) inexpensive chemical insecticides.

In modern, industrialized society, many persons consider the sight, or even the thought, of so much as a single insect to be sufficiently abhorrent as to warrant control measures. Persons who are convinced that they are under attack by arthropods even when none can be found generate considerable business for commercial pest control operators (Keh, 1983). Olkowski and Olkowski (1976) concluded that, as part of a general negative attitude toward "dirt," such people considered insects noxious. With the exceptions of butterflies, dragonflies, fireflies, lady beetles, honeybees, crickets, and mantids, all of which enjoy a favorable press (Levenson & Frankie, 1983), "the only good insect is a dead insect" for a huge proportion of North Americans and Europeans. In dealing with aesthetic EILs, the strategy of successful pest management might include development and dissemination of educational material in addition to more thorough understanding of the detailed ecology of the pests themselves.

CHAPTER

3

SAMPLING POPULATIONS

The intimate relationship between population density and EIL gives a central and practical role to accurate estimation of the numbers in a population of insects (or any other pest). Reliable estimation of population density is also an essential prerequisite to any meaningful research into factors determining distribution and abundance.

It is nearly always impossible to count accurately every single individual in a population of free-ranging animals, insects included. One therefore must devise a scheme to sample a proportion of the population, and to use such samples along with statistical techniques to infer generalizations about the population as a whole. Therefore, statements about insect populations are only as accurate as the sampling methods employed, together with the validity and applicability of the statistical analyses performed on the resulting data.

This chapter provides an overview of some of the problems and promises of various commonly used techniques for sampling insect populations, and concentrates on those techniques that applied entomologists have found especially useful for estimating pest densities and forecasting economic injury. Southwood (1978) presents a far more detailed and systematic review of the theory and practice of sampling arthropod populations.

At the outset of planning a sampling scheme, an investigator must know as much as possible about the biology of the subject population. Information on activity periods and critical thresholds of survival must be sought for temperature, humidity, light, and response to colors, lest a person be misled by the resulting data. For example, characteristically, alate aphids fly primarily during the morning and late-afternoon hours and are more likely to be collected in yellow pan traps at these times rather than at midday. Parasitic Hymenoptera rarely fly at air temperatures below 10°C, particularly when the sun is obscured, so fewer are likely to be collected in a malaise trap (Figure 3-7a, 3-8) on cloudy, cool days. The crepuscular activity of many biting Diptera is well known to scientist and layman alike. Most people also know that moths "come to lights at night," and careful study reveals species-specific differences in flight times; some of the large and showy Saturniidae are active primarily after midnight. There may be considerable individual variation in behavioral responses of insects to stimuli, and this can become especially important when traps are employed. Individual insects vary in their ability to avoid a trap, and an attractant may hold greater or lesser appeal according to an insect's age and whether or not it has mated. Statistical techniques for analysis of data often assume that individual insects behave identically or that variation in response is random, whereas neither condition may be satisfied in the real world.

THEORETICAL CONSIDERATIONS

Preliminary sampling is the first step in designing a sampling program to estimate pest density. In particular, a decision must be made at the outset of a study as to what constitutes the area of interest (the "universe"), based on the purposes and goals of the study. An intensive investigation of the effects of environmental factors on an insect species might concentrate on a single location and demands a high degree of precision, whereas forecasting pest outbreaks for growers may extend over an entire state, region, or nation, and high precision may be of secondary importance to timely information.

Before statistical inference can be legitimately used, the relationship of an insect's distribution to that of an appropriate statistical population should be determined.[1] Many statistical techniques are based on the "normal" distribution, a theoretical distribution wherein variation is random, mean and variance are independent of one another, and the curve is symmetrical with specified probabilities (e.g., 95% of variation lies within two standard deviations of the mean). To use techniques based on the normal distribution, the entomologist either assumes *a priori* that there is a reasonable correspondence between observed data and normality or he/she must determine an approximate corresponding statistical distribution whose relationship to the normal is known, whereupon the observed data can be "normalized."

1. I assume that the reader has a nodding acquaintance with statistical inference. Bishop (1967) provides a helpful introduction for the novice. In the modern world every serious student of the sciences should take a course in statistical inference.

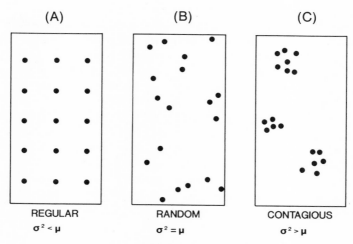

FIGURE 3-1

Examples of dispersion patterns. (A) Regular; points equidistant from one another and variance is less than mean. (B) Random; probability of any one point's location is independent of that of any other, and variance is approximately equal to mean. (C) Contagious or clumped; any location with a point increases the probability of another point's being nearby; variance exceeds mean.

Distributions of insects (or anything else) in two-dimensional space assume patterns that correspond broadly to one of three types (Figure 3-1). In a regular distribution, (Figure 3-1A), individuals are spaced evenly at (more or less) regular intervals. In a random pattern (Figure 3-1B) the occurrence of each individual is independent of the occurrence of any other; they are scattered literally at random. Observed randomness may be an artifact of sampling rather than a function of the insect's biology, in which case the pattern is more appropriately termed "pseudorandom" (Taylor, 1984). Clumped ("contagious") patterns (Figure 3-1C) are most typical of insect populations; here there is a tendency for observations to occur in small clusters. More formally, in a contagious pattern the presence of any individual increases the likelihood of another's being found nearby.

The apparent dispersion pattern may change greatly as insect density changes or as the sampling unit enlarges or shrinks, even during the course of a single field experiment (Taylor, 1984). Alate aphids often settle randomly onto plants early in a growing season, but their distribution becomes clumped soon after they commence reproduction (Horn, 1981; see also Table 3-1 and chapter appendix).

TABLE 3-1

Frequency Distribution and Expected Poisson and Negative Binomial Distributions for Green Peach Aphids Collected from 132 Collard Plants, 8 June 1979, Columbus, Ohio

Number of aphids/plant	Observed frequency	Poisson		Negative binomial	
		Probability	Expected frequency	Probability	Expected frequency
0	39	.067	9	.232	31
1	22	.183	24	.203	27
2	15	.246	32	.157	21
3	13	.220	29	.116	15
4	16	.148	20	.083	11
5	6	.079	11	.059	8
6	5			.042	6
7	3				
8	4	.053	7	.049	7
9	3			.046	6
10+	6				

$N = 132$ $\chi^2 = 182.98$ $\chi^2 = 9.39$
$\bar{x}$ (mean) = 2.69 $p < .000001$ $p = .25$
s^2 (variance) = 8.21

Brown and Cameron (1982) found that the spatial pattern of the gypsy-moth egg parasitoid *Ooencyrtus kuvanae* changed from pseudorandom to contagious as density increased. Parasitized insects may scatter more widely than those that are unparasitized (Luck & Dahlsten, 1980). Also, if the area sampled is expanded, the apparent distribution changes. If one were to sample Figure 3-1 by using a giant net covering 10 square meters, all the dots would seem clumped, because the page of a book would be a very small proportion of the total area sampled. Plant phenology changes during the growing season, and this also can alter the apparent dispersion pattern of phytophagous insects (Stanton, 1983).

Use of untransformed census data in statistical analyses based on the normal distribution is limited to censuses from randomly dispersed insects. If the insects display a regular dispersion pattern, each count (x_i) should be squared before analysis. For insects showing slight clumping, transformation of data to square roots is appropriate. Where clumping is more pronounced, transformation to logarithms— usually x_i to log $(x_i + 1)$—normalizes the data. The method of determining the apparent dispersion pattern is best illustrated by example.

(In practice, a competent statistical consultant may carry out the following procedures, and they are included here primarily for your enlightenment.)

Example: Green Peach Aphid

The green peach aphid (*Myzus persicae*) overwinters in the egg stage on deciduous trees (in Ohio) and, in May, winged females fly to crop (and other herbaceous) plants, on which they settle and commence production of parthenogenetic, wingless female offspring. In one experiment I monitored their movement onto and subsequent reproduction on collards in small plots in central Ohio. Table 3-1 shows the results from sampling 132 plants for aphids on 8 June 1979. I wished to know whether the aphids were dispersed randomly or contagiously in order to devise accurate and statistically valid sampling schemes for intensive studies of population dynamics. Sampling of the 132 plants yielded the frequency data in the table, and mean and variance were calculated. The next step was to compare the observed data with the distribution expected if the aphids had been distributed at random. The "poisson series" is a mathematical distribution describing a truncated (i.e., having high proportion of zeros) random distribution with low mean and variance independent of the mean (Southwood, 1978). Essentially one calculates mean and variance and then asks: How would aphids be distributed on 132 plants with a mean of 2.69/plant and a variance of 8.21 if their dispersion conformed to a poisson series?

Poisson probabilities are calculated from the formula

$$P(n) = e^{\bar{x}} \left(\frac{\bar{x}^n}{n!} \right) \tag{3-1}$$

where e is the base of natural logarithms (= 2.71828+), n is the expected number of aphids/plant, and x is mean number of aphids/plant. With a mean of 2.69, the probability of zero aphids per plant is

$$P(0) = e^{-2.69} \left(\frac{2.69^0}{0!} \right) = e^{-2.69} (1) = .067$$

One simply plugs 1, 2, 3, and so forth into the formula of Equation 3-1 to determine subsequent probabilities, and each probability is multiplied by 132 to arrive at the expected distribution that would result if the aphids were randomly dispersed. A chi-square value (see chapter appendix, or any decent statistics textbook) is then calculated and compared to a tabular value to arrive at the probability that the observed distribution conforms to that expected. (A statistical restriction on using this method is that an expected value cannot be less than 5.) A look at Table 3-1 suggests that the field data do not conform especially well to those generated by the poisson series, because there is an overabundance of zeros and of high numbers among the field counts. This inference is borne out by the chi-square of 182.98, a large value for a chi-square, indicating that the probability of these aphids' dispersion being random is near that of the sun's not rising tomorrow. (The chapter appendix gives an example wherein data do conform to a poisson distribution.)

The observed dispersion of aphids in Table 3-1 more likely conforms to a contagious distribution, as the variance is considerably in excess of the mean (and reproducing aphids often do exhibit a clumped distribution). The negative binomial is one (of several) statistical distributions that describe contagious dispersion patterns in which the variance is dependent upon the mean. In order to compute the expected negative binomial frequencies for 132 plants with mean = 2.69 aphids/plant and variance = 8.21, one must first estimate two additional parameters:

$$q = \frac{s^2}{\bar{x}} \tag{3-2}$$

$$k \doteq \frac{\bar{x}^2}{s^2 - \bar{x}} \tag{3-3}$$

where s^2 is the variance. (k is often termed the "index of aggregation.") These formulae provide reasonable estimates of q and k, though they are valid only for frequency data whose mean is low and includes a high proportion of zeros. (Southwood, 1978, gives more involved and more precise formulae for calculating these two parameters.)

For the data of Table 3-1, $q = 3.05$ and $k = 1.31$. One next computes the expected probabilities. The probability of no aphids is

$$P(0) = q^{-k} \tag{3-4}$$

Each subsequent probability is calculated from the previous one by

$$P(n) = P(n - 1) \left(\frac{k - (n - 1)}{n} \right) \left(\frac{1 - q}{q} \right) \tag{3-5}$$

This formula is the simplest of many, but yields only an approximation (L. C. Cole, personal communication); the more precise though complex formulae in Southwood (1978) are preferable. In the example here, $P(0) = 3.05^{-1.31} = .232$, and $P(1) = .232 \times 1.31 \times [(1-3.05)/3.05] = .203$.

As before, once a table of probabilities is calculated, each is multiplied by 132 to arrive at the expected frequencies, and a chi-square is calculated. In this example, the result is clearly within the range of acceptance; that is, chances are good that the aphids do conform to a negative binomial distribution.

It does not follow that all aphid populations on all crops conform to this model. (See chapter appendix for an example.) In fact, due to the vagaries of sampling and the insect's biology, this specific model is probably unique to green peach aphids on collards at The Ohio State University Research Farm on 8 June 1979. However, if repeated sampling over a range of densities yields data conforming to the negative binomial (although with differing k's), this strengthens the validity of the general model and its application.

The statistical properties of k and of the negative binomial are subject to lively debate (e.g., Taylor et al., 1979). Not all contagious populations conform to the negative binomial, and one must be careful not to extract unwarranted biological significance from a statistical model. The value of k is influenced by the size of the sampling unit and also by the number of samples because variance is used in the calculation of k. Southwood (1978) and others provide methods for calculation of a "common k" after several populations have been fitted to the negative binomial, though Finch et al. (1975) found "common k" to be statistically valid only over a limited range of values. It is best to calculate "common k" only for samples of equal replication, in order to stabilize the variance. Even so, k may vary so widely as to prohibit any meaningful common value. For example, Harcourt and Binns (1980) were unable to determine a stable "common k" for the alfalfa blotch leafminer after exhaustive sampling over several years.

Taylor (1961) proposed an exponential function to describe the overall relationship between variance and mean for contagious populations in the absence of a fit to the the negative binomial (or any related statistical distribution). "Taylor's Power Law" (as it is now known) is

$$s^2 = a\bar{x}^b \tag{3-6}$$

where $a = s^2$ when $x = 1$, and b is the slope of plot log s^2 on log x. Taylor (1984 and references therein) suggested that a varies with the sampling technique while b seems to assume a typical constant value for each species. As such, it may be a more appropriate "index of aggregation" than k of the negative binomial. Taylor *et al.* (1978) presented a convincing case supporting the ubiquity of the Power Law with an analysis of 200,000 samples representing 102 animal species, all of whose statistical distributions were best described by Equation 3-6. According to Ruesink (1980), k of the negative binomial is related to the Power Law by

$$ k = \frac{\bar{x}}{a\bar{x}^{(b-1)} - 1} \tag{3-7} $$

where a and b are constants as in equation 3-6. Moreover, if $a = b = 1$, then variance equals mean and the data can be fitted to the poisson distribution.

Lloyd (1967) and Iwao (1968) devised a "mean crowding index" as another measure of aggregation when the negative binomial is found to be inadequate:

$$ \overset{*}{M} = \bar{x} + \left(\frac{s^2}{\bar{x}} - 1 \right) \tag{3-8} $$

This has been used in numerous studies, examples being those of Bechinski and Pedigo (1981) on predatory stinkbugs in soybeans, and Shepherd (1985) and Shepherd *et al.* (1984) on the Douglas-fir tussock moth. Mean crowding is a useful and simple index of aggregation, though it is subject to some statistical difficulties (see "sequential sampling," below).

The negative binomial remains an adequate description of the dispersion patterns of many species (e.g., Guppy & Harcourt, 1970, 1973), though should be used with caution. Once it has been determined that observed data from preliminary sampling conform to a contagious distribution, the following are in order.

TRANSFORMATION

As mentioned earlier, transformation of data to $\log x_i$ or $\log (x_i + 1)$ is appropriate for contagious populations and is necessary to normalize the distribution and to separate the dependency of variance on mean. For populations that conform to Taylor's Power Law, transformation to $x_i^{1-\frac{1}{a}b}$ is appropriate (Ruesink, 1980). Once transformation is completed, statistical analyses based on the normal distribution (t-test or analysis of variance with accompanying mean separation techniques) can be applied legitimately to the data.

LEVEL OF ACCURACY

At the outset of this chapter, I stated that the level of accuracy is dependent on the purposes of the study. Coarse but inexpensive sampling techniques are often adequate for determining gross changes in pest density, and an error of 25% of the mean may be adequate in such cases. Greater precision is required in intensive population studies such as life table censuses intended to serve as bases for population models (Chapter 4). An error of 10% is desirable in such studies. Obviously, the greater the level of accuracy one desires, the more extensive must be the sampling. Rojas (1964) related level of accuracy to the negative binomial via the following formula:

$$n = \frac{\frac{1}{\bar{x}} + \frac{1}{k}}{A^2} \tag{3-9}$$

where A is the desired level of accuracy (= error), and n is the number of samples necessary. For the aphid population study, an accuracy level of 10% was desired, so

$$n = \frac{\frac{1}{2.69} + \frac{1}{1.31}}{(.10)^2} = 113$$

Therefore, 113 plants must be sampled on each sampling date to obtain an error of 10% around the true population mean. (This was a most gratifying result, in that 132 plants had been sampled initially, so that data from 8 June 1979 could be legitimately incorporated into the intensive population study.)

If the negative binomial has been found inadequate, the level of accuracy can be related to the constants from Taylor's Power Law:

$$n = \frac{a\bar{x}^b}{A^2} \tag{3-10}$$

SEQUENTIAL SAMPLING

The negative binomial is helpful in devising a sequential sampling scheme to determine the statistical validity of a particular EIL. Sequential sampling was developed for quality control in the munitions industry during World War II and has won wide acceptance in recent decades as a means of providing a cost-effective technique for population estimation usable by pest management scouts in the field.

The first step in formulating a sequential sampling plan is to set upper and lower limits for the EIL. For this exercise, I again used the data from the green peach aphids. The EIL is subject to the usual arbitrary factors (Chapter 2); for this exercise, I assumed that a density of two or fewer aphids per plant was never economically important, and that a density of five or more was invariably damaging. (These EILs were chosen for the purpose of this exercise alone and actually are somewhat lower than generally accepted EILs for aphids on vegetable crops.) Between two and five aphids per plant lay a region of uncertainty: aphids might cause economically important damage, or they might not. The next step was to choose probabilities for "Type I" and "Type II" errors. (These are briefly discussed in the appendix to this chapter; for more detail, the reader is advised to consult a reference on statistical methods.) If the data have been shown to conform to a negative binomial, as above, one next calculates

$$P_1 = \frac{\bar{x}_1}{k} \tag{3-11a}$$

$$P_2 = \frac{\bar{x}_2}{k} \tag{3-11b}$$

where $\bar{x}_1$ and $\bar{x}_2$ are the minimum and maximum EILs previously chosen (two and five aphids/plant).

"Sequential sampling lines" are next determined by

$$m = k \frac{\log\left(\frac{1 + P_2}{1 + P_1}\right)}{\log\left(\frac{P_2 (1 + P_1)}{P_1 (1 + P_2)}\right)} \tag{3-12}$$

$$y_0 = \frac{\log\left(\frac{\alpha}{1 + \beta}\right)}{\log\left(\frac{P_2 (1 + P_1)}{P_1 (1 + P_2)}\right)} \tag{3-13a}$$

$$y_1 = \frac{\log\left(\frac{1 - \alpha}{\beta}\right)}{\log\left(\frac{P_2 (1 + P_1)}{P_1 (1 + P_2)}\right)} \tag{3-13b}$$

where α is the probability of Type I error, β is the probability of Type II error, m is the slope of sequential sampling lines, and y_0 and y_1 are the intercepts of sequential sampling lines. (One uses natural logarithms in making these calculations.) In our example, $P_1 = 1.53$, $P_2 = 3.82$, $m = 3.12$, $y_0 = 8.12$, $y_1 = 8.12$. Armed with all this, we calculate the sequential sampling lines with the familiar formula for a straight line

$$y = mx + y_0 \qquad\qquad \text{(3-14a)}$$

$$y = mx + y_1 \qquad\qquad \text{(3-14b)}$$

The lines are then plotted (Figure 3-2). A pest management scout may now take this graph along on a visit to the field, and, as he or she

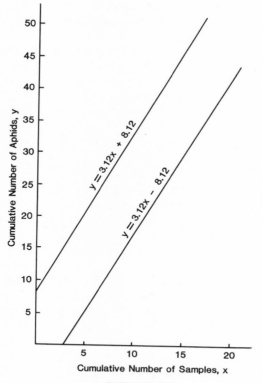

FIGURE 3-2

Sequential sampling lines for green peach aphids on collards (example discussed in text; data of Table 3-1).

commences to sample plants, the cumulative number of aphids (y) is plotted against the cumulative number of plants sampled (x) until the plotted line enters either of the areas of decision. Figure 3-3 depicts an example based upon the calculations above. A scout entering Field #1 counted, on successive plants, 7, 2, 3, 4, and 6 aphids. Sampling ceased as soon as the cumulative plot entered the area "Above EIL," and the scout could then state with 90% certainty that the aphid population in this field exceeded the EIL. In Field #2, the scout found 3, 2, 0, 0, 4, 2, 0, 3, 1, and 0 aphids on 10 successive plants, whereupon the "Below EIL" area was entered and sampling ceased. Again, a decision (that this time the EIL was not exceeded) can be made with 90% assurance that it is correct. It can also be stated (with 90% assurance) for any field wherein the first plant sampled has 12 or more aphids that the EIL is exceeded. (My students have a hard time believing that a single sample has statistical validity. It does, though a prudent pest management scout would do well to sample a few other areas of the field anyway; one must recall the caution against placing too much biological significance on a statistical model.)

A scout might find a mean of three aphids/plant with very little variation, so that the cumulative total would continue to plot within the area of uncertainty despite many successive samples. In practice, therefore, an arbitrary cutoff is established (20 samples, in this example). If the accumulated total has not emerged from the area of uncertainty after 20 samples, the scout simply reports the situation as "uncertain," and might note this particular location as a candidate for an early revisit.

Onsager (1976) and Southwood (1978) discuss further the details of sequential sampling and provide formulae for computing decision lines for populations that conform to poisson or binomial distributions. Sequential sampling decision lines can also be generated from regression of log variance on log mean after Taylor's Power Law (Green, 1970) or from the mean crowding index (Iwao, 1975), though Nyrop and Simmons (1984) noted that Iwao's method compounded error rates beyond the probabilities initially chosen for Type I and Type II errors. Grothusen (1984) developed a program using a hand-held calculator to adjust sequential sampling lines in the field according to prevailing conditions. Pedigo and van Schaik (1984) suggested the use of sequential sampling over time (rather than space) to detect the timing of green cloverworm outbreaks. The method might be applicable to such data as light trap surveys, though this has yet to be attempted. Numerous studies have demonstrated that sequential sampling is an improvement in accuracy, efficiency, and tedium over other techniques because a fixed sample size is likely to be imprecise at low

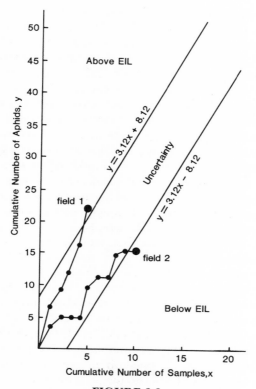

FIGURE 3-3

Sequential sampling lines for green peach aphids on collards (lines and equations same as in Figure 3-2). Areas of decision added along with examples of cumulative counts from each of two hypothetical fields. In Field #1, EIL is exceeded; in Field #2, it is not (see text).

pest density and inefficient at high pest density. For example, Bellinger *et al.* (1981) showed time savings of 30% to 68% with accuracy of 82% to 91% in a sequential sampling scheme.

Pieters's (1978) bibliography contains numerous references attesting to the value of sequential sampling in practical and theoretical studies of pest populations. Even more precise methodology is in development. For instance, Hoy *et al.* (1983) criticized sequential sampling because its use may omit large parts of a field. They developed a pattern of sampling throughout a (cabbage) field, varying the number of plants sampled at each station. Their method needs refinement, but may be suitable for sampling insect populations displaying a high variance and uneven distribution in the field.

PRACTICAL CONSIDERATIONS

Morris (1960) and others have suggested that valid sampling techniques should take into account the following considerations.

1. All units within the habitat of interest (the "universe") must have an equal chance of selection. Random selection satisfies this condition, though an investigator must guard against any systematic but unconscious bias. Bias can be introduced when an investigator simply chooses a sampling unit "at random" independently of any objective means of determining randomness; this is haphazard sampling, and not truly random. When selecting units of a field for sampling, it is best to use an independent, truly random means of selecting sampling units—dice, playing cards, or (preferably) a random-number generator. (The final two digits in a column of telephone numbers from the white pages of a telephone directory in a large city, e.g., Columbus, Ohio, are usually an adequate substitute for a random-number generator.)

2. The proportion of the population within each sampling unit must be reasonably constant. This may be a constraint on #1 above. Populations of many phytophagous insects are lower on plants at or near the edge of a field when compared to numbers near the center. This so-called "edge effect" may adversely influence the accuracy of sampling. Preliminary sampling may reveal the prevalence of edge effect, which is well documented in the literature for some pests. Stratified sampling may be necessary to counter diurnal or seasonal movement of insects in three-dimensional habitats, such as forest trees (Luck & Dahlsten, 1980; Coulson et al., 1976).

3. The sampling unit must be readily convertible to unit areas, for legitimate comparison with other studies on the same insect species. This is further elaborated in discussion of relative estimates below.

4. Either the sampling unit must be stable in the field, or changes must be easily measured. If one samples row crops, for example, the biomass of the plants normally increases during the growing season, and this affects the apparency of the area sampled. "Insects per plant" may mean two quite different measurements early and late in the season.

5. The sampling unit must be easily delineated in the field. Ideally, anyone should be able to obtain acceptable samples after minimal training.

6. There must be a reasonable balance between sample variance and cost of sampling. This may be estimated by substituting into Equation 3–9 the cost per sample for various sampling techniques. For

example, in June it takes me 5 minutes to sample aphids from an entire collard plant and I need 113 samples for 90% accuracy (p. 43). If the hourly wage were $5/hour, sampling 113 plants would cost $47.08. To detect gross changes in aphid density might require accuracy of only 75%. This would require sampling only 18 plants at a cost of $7.50, a savings of $39.58 from a reduction of 15% in accuracy. Great accuracy at horrendous cost may be necessary to detect extremely sparse infestations, as is necessary in enforcement of quarantines (Chapter 11).

Sampling Interval

Insect development is closely related to ambient temperature, and this must be taken into account in designing routine sampling programs. Insect populations change faster in warm than in cold weather. It is therefore more appropriate to base sampling intervals on "degree-days" than on calendar days. Pruess (1983) and Higley et al. (1986) reviewed advantages and disadvantages of the various uses of the degree-day concept in entomological studies.

The first step is to determine the developmental and activity threshold temperature for the insect under study. This varies, though for many temperate-zone species it is around 10°C. The straight-line relationship between temperature and development is often extrapolated to zero to calculate the threshold, though this may lead to inaccuracy because the relationship may be curvilinear near the extremes for development (Stinner et al., 1974).

After the relationship between temperature and development is determined, several methods may be used to calculate degree-days. The most accurate is integration of temperatures over an entire 24-hour day, though a sine-wave function is also adequate. Both involve use of large databases and rather sophisticated computer simulations. A crude but effective technique is simply to calculate the mean of the daily maximum and minimum temperatures and subtract the developmental threshold. For example, yesterday the National Weather Service reported a maximum of 24° and a minimum of 8°, for a mean of 16°. The developmental threshold for alfalfa weevil is 9°C (Armbrust et al., 1980), and, thus, yesterday accumulated $16 - 9 = 7$ degree-days. If the identical temperatures prevailed for a week, 49 degree-days would accumulate.

For convenience and standardization it is best to use the data provided by the U.S. National Weather Service (or a similar network), although these data may not be accurate for estimating development of an insect species unexposed to standard weather conditions. Overwintering pupae beneath 5 centimeters of soil are subjected to higher winter and lower spring temperatures than are insects (or humans) exposed to ambient air.

Specific Sampling Techniques

The discussion to follow concentrates on some of the more commonly used (and often abused) sampling techniques. I have not attempted exhaustive coverage, details of which are readily available (see Lewis & Taylor, 1967, and Southwood, 1978). Modifications of standard (and unique) sampling techniques are legion, limited only by the ingenuity of the investigator and careful consideration of Morris's six criteria, discussed above.

ABSOLUTE ESTIMATES

Absolute estimates of population density are those in which either an estimate is made of the actual population size (as in capture–recapture) or the sample is related directly to standard units (as insects/square meter). Absolute estimates are particularly critical to intensive population studies, such as life table analyses (Chapter 4) where a reasonably low variance is desired.

Capture–recapture techniques provide an absolute estimate of the total number within a population. A sample of the population is captured, marked in a recognizable and indelible fashion, and released back into the population. A subsequent sample is drawn from the same population, and the density is then computed from the "Lincoln Index" (Lincoln, 1930)

$$N = \frac{Mn}{m} \qquad (3\text{-}15)$$

where M is the total number marked originally and released, n is the total number captured on next occasion of sampling, and m is the number of marked individuals recaptured.

Among insect pests, larger and more sedentary species (such as domestic cockroaches) are most amenable to this approach. The accuracy of the estimate depends on the extent to which the following assumptions are satisfied.

1. The mark is stable and has no deleterious effect on the insect. This should be obvious, though a surprising number of researchers

persist in using paints or dyes that contain toxic chemicals. Preliminary studies should be undertaken in the laboratory to verify both harmlessness and persistence of the mark. Artists' acrylic paint seems suitable in most cases, though it does not adhere to oily insects, such as cockroaches. Fluorescent powder also gives favorable results.

2. The marked individuals redistribute themselves evenly within the population after release. This is extremely difficult to verify, though an investigator can increase the chances of even redistribution by marking and releasing insects as encountered, rather than by mass-releasing all marked individuals at a single location.

3. No immigration, emigration, birth, or death occurs between initial marking and recapture. This is also nearly impossible to verify, though the effects of such demographic changes can be minimized by confining the study to times and places for which one already expects that mass movement is unlikely, and by recapturing as soon as practicable after initial release—the sooner, the better, giving sufficient time for the insects to redistribute.

4. Equal "catchability" of marked and unmarked insects is assumed. If a trap of some kind is used, one must take care that marked insects are not preferentially attracted or repelled. (This problem is more serious in studies of birds or mammals, whose populations may contain both "trap-shy" and "trap-happy" individuals.) Recapture rates may vary according to placement of traps (Elkinton & Cardé, 1980). The greatest problem in most capture–recapture studies of insects is a low rate of recapture of marked individuals (Southwood, 1978); this can be ameliorated only by initially marking and releasing astronomical numbers.

Some of these potential difficulties can be solved by repeated marking and subsequent sampling, resulting in a series of successive estimates of density. Marks can be varied sufficiently to allow recognition of each sampling date, which aids in determining the extent to which conditions (3) and (4) are satisfied. Such extended sampling is also useful in determining survivorship for use in life tables (Chapter 4). Begon (1979) is a helpful general reference for capture–recapture techniques.

Absolute estimates can also be had by sampling a unit area, either by counting directly or by an extraction technique. Suction traps provide accurate estimates of flying insects per volume of air sampled (Lewis & Taylor, 1967). Quadrats of 0.1 square meter (or any other size) can be placed directly over plant cover, and the insects within are sampled by a variety of techniques. In sampling row crops, an accurate estimate of larger and relatively slow-moving species (such as caterpillars) can be obtained by placing light-colored cloth beside plants in the row, then shaking the plants vigorously and counting the insects per meter of row (Pedigo *et al.*, 1972).

An extraction technique that has gained widespread acceptance is a berlese (or tullgren) funnel for removal of insects from a measured volume of soil, litter, or vegetation (Figure 3-4). A sample from the field is placed onto a screen atop the funnel, and an overhead source of heat slowly warms and dries the contents. Arthropods descend downward into a collecting jar. Care must be taken that the heat is not too intense, lest the insects be cooked to death before they wriggle through the screen. A surface temperature of about 40°C is usually enough, and this can be supplied by a 60-watt incandescent bulb. Too much material should not be loaded in at once, and the funnel should be large enough to completely dry a sample in 24 to 36 hours. Obviously a berlese funnel extracts only active stages. Efficiency of berlese funnels

FIGURE 3-4
Berlese funnel. Heat source at top drives insects through screen within the funnel and into collecting jar below.

FIGURE 3-5

Vacuum insect net. Motor-driven fan creates airflow through large flexible hose. Insects are drawn into net within the large white "collecting head."

should be checked periodically before and during a study by dumping a known number of insects into the funnel along with uninfested soil (or vegetation) and counting the number of insects recovered.

An alternative to the berlese funnel is extraction of eggs and pupae from soil or vegetation via flotation or sifting. Washing soil samples through a series of sieves and then floating eggs or pupae on a dilute salt solution provides a satisfactory estimate of population density, though it is extremely time-consuming.

For the more mechanically inclined entomologist, vacuum insect nets are available for sampling per unit of vegetation-covered surface. Dietrick (1961) introduced the most popular model, the D-vac (Figure 3-5), though others exist (e.g., Summers *et al.*, 1984). A vacuum insect net consists of a gasoline-powered engine driving a fan that sucks insects into a mesh bag whose contents may then be fumigated, frozen, or otherwise held for later sorting and counting. Vacuum collection has been demonstrated to be very efficient (Callahan *et al.*, 1966), though the entrance to the collecting head must be placed rather tightly over the area to be sampled, much as in vacuuming a rug on a floor. Even so, particularly when the soil surface is damp, larger and heavier insects (such as beetles and caterpillars) may be left behind after vacuuming (Lewis, 1977). Vacuum insect nets, like any mechanical contraption, may develop problems when afield, and it is well for the researcher to have a small toolbox available, and to know something about internal-combustion engines.

RELATIVE ESTIMATES

Relative estimates express population density as a function of effort rather than of area or volume. Examples are net sweeps or the many sorts of insect traps. Estimates of "intensity," relating insect numbers to units of habitat (such as aphids/plant) are also relative estimates, as are population indices based on censusing results of insect activity (frass, exuviae, feeding scars, etc.). Relative estimates are often easier to obtain in the field, though many relative sampling techniques yield a higher variance than does a corresponding absolute technique. Moreover, relative estimates are difficult to interpret except by comparison with other relative estimates. Some of these difficulties can be addressed by conversion of relative to absolute estimates.

The sweep net is a time-honored though crude technique for quickly deriving relative estimates of insect populations. The imprecision of sweep netting has been known for many years (DeLong, 1932; Fenton & Howell, 1957) and results from such variables as insect activity (Cothran & Summers, 1972), growth stage of plants (Horn, 1971), and differences among individual sweepers using the technique (Fleischer *et al.*, 1982). Lively debate has centered around determining what constitutes a net-sweep. A 180° arc results in a lower variance than a 90° pendulum-type swing for some insects but not others. The sweep net remains a widely applied technique for collecting over a large area quickly, and for measuring gross changes in density such as may be important in scouting fields for pest management. However, sweeping is of questionable utility in more intensive studies of insect population dynamics.

Density estimates based on sweeping are convertible to absolute estimates if the variance is not especially large and if a significant proportion of the population is not being missed systematically. Fleischer *et al.* (1982) found that the number of adult potato leafhoppers per 0.9 square meter of alfalfa could be estimated by the formula $N = 18.5 \times S$, where S is the number of leafhoppers per sweep. The formula was determined by regression of vacuum net samples on sweep samples, each of which had reasonable stability of variance. On the other hand, Cothran and Summers (1972) found that a sweep net failed to collect a significant proportion of early-instar larvae of the Egyptian alfalfa weevil, and thus greatly underestimated their density early in the growing season, when knowledge of weevil numbers was particularly critical to growers.

An absolute estimate of density also may be had from "removal sweeping," a technique wherein a series of successive passes is made over the same area, and the population density is extrapolated from successively declining counts. Figure 3-6 shows an example in which I

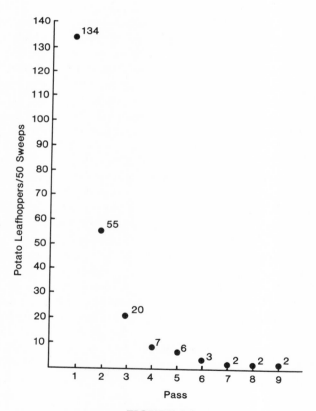

FIGURE 3-6
Results of sweep netting for potato leafhoppers in successive 50-sweep samples over same 20-square-meter area of alfalfa, Columbus, Ohio, 6 July 1982. Estimated total leafhoppers (9 passes) = 231.

repeated 50-sweep samples over the same area of alfalfa. On each early pass the number of potato leafhoppers was less than half that of the preceding sample. Once the number collected was negligible (= 2), I simply totalled the cumulative number collected. Ideally, the same person sweeps on each pass, because people can differ markedly in sweeping technique. Also, the method works better for sedentary species (e.g., alfalfa weevil or apterous aphids) than for highly active ones (e.g., potato leafhopper). An absolute estimate can then be determined by careful measurement of the area swept.

Relative estimates of density can be obtained by traps of many sorts. Figure 3-7 depicts some commonly used insect traps. Traps may be active, baited with an attractant, or passive, merely intercepting insects moving through the environment. Both baited and unbaited traps depend on insect activity for their effectiveness, and are actually monitoring that activity rather than population density per se. The behavior of the insect species in question should be investigated as thoroughly as possible before valid conclusions about density can be drawn based on the results from trapping. For example, Chiverton (1984) found that pitfall traps contained more carabid beetles from insecticide-treated plots than from untreated controls, apparently because the insecticide had killed off many prey species and the starved beetles were more actively searching for prey. Efficiency of trapping is especially critical in detection of low-density infestations necessary in administration of quarantine programs.

Malaise traps (Figure 3-7A) are commonly employed to collect flying insects, such as biting flies or parasitic wasps. Malaise traps may be passive or active; baited with a source of carbon dioxide, they attract biting Diptera in much greater quantity than when unbaited. Ambient temperature is also important for efficient operation (of these and most other insect traps). Figure 3-8 illustrates the influence of temperature on malaise-trap catches of Ichneumonidae in a suburban backyard.

Light traps are also dependent upon temperature and apparently on cloud cover for accurate samples of many species; the relative ineffectiveness of light trapping on clear moonlit nights is well-known to collectors of moths. There are also differences in attractiveness of males versus females to light. Rather little is known of the behavioral aspects of insects' response to lights. Different species are evidently attracted over varying distances, which can confound the interpretation of light trap results.

Similar considerations apply to efficient use of the other sorts of traps illustrated. Pheromone trap catches may vary with the position of the trap, height being particularly important. As will be seen (Chapter 8), some sex pheromones are very complex isomeric blends not easily mimicked by synthetic compounds. Again, more research is needed on insect behavior, particularly close-range orientation to (or away from) the trap.

For additional information concerning the details of trapping insects, Southwood (1978) is a particularly useful reference.

APPENDIX

Here is another example of fitting an observed to a theoretical distribution and drawing sequential sampling lines. As in the earlier instance, the insect involved is the green peach aphid. In this case, winged aphids settled upon collard plants in small plots at Davis, California. I sampled 132 collards on 13 October 1982 and found the frequency distribution that is presented along with mean and variance in Table 3-2.

The question was again posed: Is this distribution random? That is, can it be described by a poisson distribution with mean of 1.15 and variance of 2.02? More formally, two hypotheses are posed: the "null hypothesis," $H_0 =$ there is no difference between the observed distribution and poisson series, and the "alternate hypothesis," $H_a =$ there is a difference between the observed distribution and the expected poisson series.

As before, probabilities are calculated for the appropriate frequency classes (by Equation 3-1), and each is multiplied by 132 to calculate the expected frequencies. A chi-square is calculated from the usual formula

$$\chi^2 = \sum \frac{(O_i - E_i)^2}{E_i} \tag{3-16}$$

where O_i is the observed number in each class, and E_i is the expected number in each class. One must next choose a probability, or "level of acceptance." Informally, this is the chance of being wrong. Formally, this is called "Type I error," the probability of accepting H_a when H_0 is true. By convention, in most entomological studies, Type I error is usually set at 5% (though there is nothing to prevent selection of a larger or smaller value). One next secures a table of chi-square values (given in an appendix to most statistics textbooks) and compares the calculated chi-square (here, 7.23) with the tabular value for $p = .95$ and four degrees of freedom (one fewer than the number of frequency classes; for negative binomial, degrees of freedom equal the number of classes minus two). If this value is greater than that calculated (and it is), one may accept H_0 as true 95% of the time. The data do indeed conform to a poisson series, and one may assume that the aphids are distributed randomly.

(B)

(A)

(D)

(C)

FIGURE 3-7

Frequently used trapping techniques for sampling insects of economic importance. (A) Malaise trap. Flying insects hit the central baffle and move upward into the collecting jar. (B) Ultraviolet light trap ("blacklight"). Insects attracted to light hit metal baffles and fall into chamber beneath. (C) Pitfall trap. Ground-dwelling insects blunder into can and drown in ethylene glycol. (D) Pheromone trap. Male moths (here, *Pseudaletia unipunctata*, the armyworm) are attracted to synthetic sex pheromone and are trapped on sticky surface.

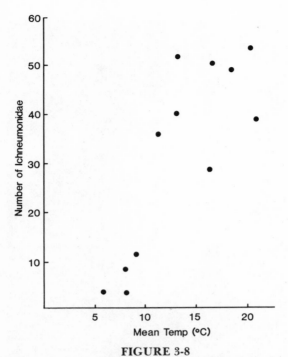

FIGURE 3-8

Relationship between mean ambient temperature and numbers of Ichneumonidae collected per day in passive malaise trap, Davis, California, October–November, 1982.

Sequential Sampling for the Poisson

When data conform to a poisson, sequential sampling lines are calculated from the following formulae:

$$m = \frac{(\bar{x}_2 - \bar{x}_1)}{\log \bar{x}_2 - \log \bar{x}_1} \tag{3-17}$$

$$y_0 = \frac{\log\left(\dfrac{1-\alpha}{\beta}\right)}{\log \bar{x}_2 - \log \bar{x}_1} \tag{3-18a} \qquad y_1 = \frac{\log\left(\dfrac{\beta}{1-\alpha}\right)}{\log \bar{x}_1 - \log \bar{x}_1} \tag{3-18b}$$

(The terms are as defined earlier. "Type II error" (β) is the probability of accepting H_0 when H_a is true.)

Again we choose $\alpha = \beta = .10$, enter the values for sequential sampling lines, and plot them as in Figure 3-9.

TABLE 3-2

Frequency Distribution and Expected Poisson Distribution for Green Peach
Aphids Collected from 132 Collard Plants, 13 October 1982, Davis, California

Number of aphids/plant	Observed frequency	Poisson probability	Expected frequency
0	49	.316	42
1	47	.364	48
2	17	.209	27
3	12	.080	11
4 or more	5	.031	4

$\bar{x}$ (mean) = 1.15 $\chi^2 = 7.23$

s^2 (variance) = 2.02 $p = .13$

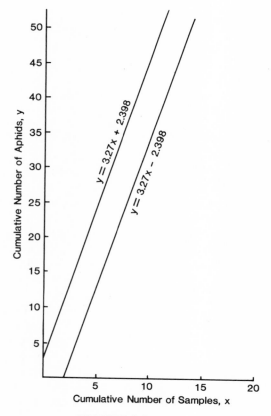

FIGURE 3-9

Sequential sampling lines for green peach aphids on collards (example discussed in text; data of Table 3-2).

CHAPTER

4

SINGLE-SPECIES POPULATIONS

Ecological interactions determining the distribution and abundance of insects in the field are enormously complex. Facing this potentially bewildering complexity, ecologists often resort to simplification of reality for insight into ecological processes. Often the resulting models are admittedly oversimplified, yet from conceptually simple models a substantial array of outputs can be produced to illustrate general principles of insect population management. This approach is followed in the present chapter, initially introducing excessively simplified models, then adding successively greater levels of complexity. For simplicity, single-species populations are considered in this chapter, though no insect (or any other) population exists completely apart from those of other species. Insect genetics and adaptations to inclement conditions are also discussed.

Formally defined, a biological population is a group of individuals actually or potentially capable of interbreeding and thereby producing fully fertile offspring. In practice, both temporal and spatial boundaries of natural populations are often determined arbitrarily. For instance, one may be primarily interested in the population of alfalfa weevils in a 40-hectare field in Indiana, and may distinguish this population from one of alfalfa weevils in Ohio, or even from one in an adjoining field across a highway 10 meters wide. As noted in Chapter 3, alfalfa weevils themselves might not recognize arbitrary boundaries, and weevils in Indiana can, and do, interbreed freely with those in Illinois, Ohio, and Michigan. However, there is a greater probability that weevils within a single field form an interbreeding population than that they interbreed with weevils from far distant fields. Entomologists rarely measure the actual genetic interactions within and between populations, but rather they assume that members of a single species comprise an interbreeding population. The assumption is not always valid, though determining the actual limits of an interbreeding population in the field is an exceedingly difficult task. It is sometimes satisfactory to broaden the operational definition of a population to include a guild of closely related species that, from the standpoint of pest management, cause very similar damage and can be managed using a comprehensive technique. Legitimately, one might therefore speak of the "population" of beetles and moths infesting a grain-storage facility; several species are involved but all infest the grain and all may be managed with a single fumigation treatment.

The population models developed in this chapter apply primarily to single-species populations. Simple population models designate the density of an insect population by a single variable (usually N), which conveys the impression that all members of a population are identical. Of course, they are not any more identical than are all members of a human population. Insects vary according to age (see section on life tables, below), behavior, morphology, genotype, or any of numerous other characteristics. It is therefore more appropriate to consider N to represent a range of individuals assumed to be identical only for purposes of study and analysis. Computer simulations based upon mathematical models often analyze population events as though all individuals were identical, but one must be continually aware that conformity is much more likely in mass-produced computer circuitry programmed by electronic instructions than among individual creatures programmed by heterogeneous DNA in a patchy and variable environment.

In the discussions to follow, population systems are introduced in simplified form, and the following general characteristics (Ehrlich & Birch, 1967; Ehrlich *et al.*, 1975) are postponed for later discussion (or ignored).

1. Populations and their effective environments are continuously changing, and a model describing population dynamics at one time and place may not adequately represent events at another time and place. (This is illustrated by the dispersion patterns of aphids described in Chapter 3. Tables 3-1 and 3-2 each show a frequency distribution of aphids on collards, yet, as was demonstrated, these populations are quite different from one another.)

2. For practical reasons it is necessary to delineate and investigate local populations, though corroborating samples eventually should be obtained from the species' entire geographic range, insofar as this is practicable.

3. Variation within a local population may well equal or exceed variation among adjacent or distant populations of the same species.

4. Immigration and gene flow may be distinctly different processes. One must not assume that changes in gene frequency necessarily follow after immigration.

SIMPLE POPULATION MODELS

Exponential Model

The simplest population model is one that regards density (N) as a simple function of the number of births (B) and deaths (D):

$$N = B - D \qquad (4\text{-}1)$$

(over any time period). However, births and deaths occur only in a preexisting population, so one must add an initial population density N_0 to the model. Then N_t, the density after any time period, t, becomes

$$N_t = N_0 + B_t - D_t \qquad (4\text{-}2)$$

where B_t and D_t are the numbers of births and deaths, respectively, during the time interval t.

It is customary to express births and deaths not as whole numbers from census data, but as rates, wherein the birthrate, b, and the death rate, d, are each the number of births or deaths divided by population density. (That is, $b = B_t/N$.) Equation 4-2 thus becomes

$$N_t = N_0 + N_0(b - d) \qquad (4\text{-}3)$$

and, for any time interval,

$$N_{t+1} = N_t + N_t(b - d) \qquad (4\text{-}4)$$

This is a difference equation describing exponential population growth, and is a simple mathematical expression of the familiar argument used to calculate the potentially astronomical densities of creatures, from houseflies to elephants, that would fill the visible universe in the absence of any constraints to population growth (Malthus, 1798). The model is oversimplified and unrealistic, yet it adequately approximates changes in density following initial colonization of a favorable environment, such as the one that aphids find on cabbages or that weeds find in a fallow field. Figure 4-1 illustrates a range of

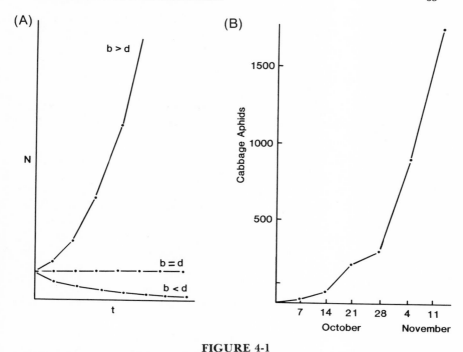

FIGURE 4-1

Exponential population growth. (A) Results of simulations based on Equation 4-4; b = birthrate, d = death rate. (Simulations run on NCF Decisionmate V.) (B) Population growth of cabbage aphids on 10 collard plants, Davis, California, 1982.

outputs emanating from this model, given constant rates of birth and death.

As noted, Equations 4-2, 4-3, and 4-4 are difference equations; the unknown (N) appears on each side, and these equations therefore have no formal solution for N, except over the time interval in question. Difference equations are useful for simulating population dynamics, especially because population events are discrete (rather than continuous) phenomena. The electronic circuitry within a digital computer is also discrete. However, the mathematics of higher-order difference equations becomes exceedingly cumbersome. For this and other reasons, population models are often expressed in differential form. A differential equation describes a continuous function, which many insect population curves approximate as densities increase and intervals between successive samples shorten. The exponential model of Equation 4-4 can be expressed in differential form as

$$N_t = N_0 e^{(b - d)t} \qquad (4\text{-}5a)$$

where e is the base of natural logarithms, and t is the time interval.

The quantity $(b - d)$ is usually expressed as r, the instantaneous rate of increase:

$$N_t - N_0 e^{rt} \qquad (4\text{-}5b)$$

(This form of the exponential growth model may be familiar to readers who are acquainted with calculus. As an exercise you can work out the algebraic proof that the first derivative of Equation 4-5a is simply the limit of Equation 4-4 as the time interval approaches zero.)

Logistic Model

The exponential model above is of limited practical use in describing population events except under unusual circumstances over short time intervals. Environmental constraints prevent the world from filling with aphids, or elephants, or humans. (Growth of the human population has approximated the exponential during the past few centuries, but it won't for very much longer.) Verhulst (1838), Lotka (1920), and Volterra (1926) share credit for development of a simple mathematical model recognizing the tendency of populations to be regulated about an upper limit set by both extrinsic and intrinsic factors. In the simplest form of their "logistic" model, K (the environmental carrying capacity) is incorporated into Equation 4-4 to act as a brake on population growth:

$$N_{t+1} - N_t = N_t (b - d)\left(\frac{K - N_t}{K}\right) \qquad (4\text{-}6)$$

This equation gives the familiar and intuitively satisfying sigmoid curve (Figure 4-2A, see also Figure 2-1). In the ecological literature, including most elementary texts, this model is more often found in its differential form:

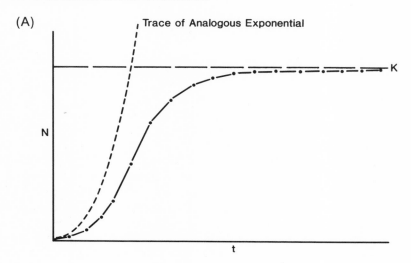

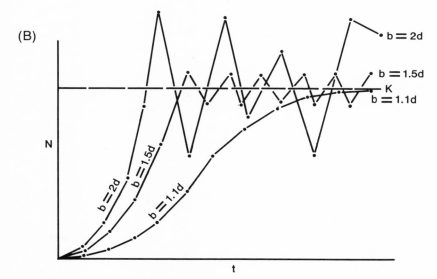

FIGURE 4-2

Logistic population growth, with modifications. (A) "Classical" logistic curve (Lotka–Volterra model), Equation 4-7. (B) Results of simulations based on Equation 4-6, changing only relationship between birthrate (*b*) and death rate (*d*). The higher the birthrate, the more chaotic and unpredictable are fluctuations about *K*. (Simulations run on NCR Decisionmate V.)

$$\frac{dN}{dt} = rN \left(\frac{K - N}{K}\right) \tag{4-7a}$$

or, in integral form:

$$N = \frac{K}{1 + e^{-rt}} \tag{4-7b}$$

where e and r are as in Equation 4-5.

Despite its simplicity, this model is capable of a large array of outputs. Figure 4-2B shows examples. Most importantly, if b increases relative to d (characterizing a population with a high intrinsic rate of increase), there is a tendency for greater oscillations about K, along with greater instability. Berryman (1981) and May (1975) discussed this in greater detail, though the general lesson is that insect populations with short generation time and high fecundity may fluctuate unpredictability and may exhibit spectacular local instability. Such populations also reach the EIL much more quickly than do those with lower r.

Logistic with a Minimum

Further realism can be added to the logistic model by considering a minimum population, a density below which the population is unable to maintain itself. This is added to the logistic in a term creating an unstable equilibrium at an "extinction threshold," E, as follows:

$$N_{t+1} - N_t = N_t (b - d) \left(\frac{K - N_t}{K}\right) \left(\frac{N_t - E}{N_t}\right) \tag{4-8}$$

If E exceeds N_t, the right side of the equation becomes subtractive, and the population declines to extinction. This is illustrated in Figure 4-3. Trivially, the population declines if the initial N_0 is less than E. This models events such as the failure of an agent of biological control to become established following initial release. Also, the density of a population with a very high r (and thus prone to chaotic fluctuation) might enter the "extinction zone" (N_t less than E) and decline to extinction (Figures 4-3A and B).

The characteristics of this model may be portrayed by an analogy of a ball rolling on a surface shaped like that in Figure 4-3C. The ball placed slightly to the right of point E rolls down toward K, where it settles, after a series of oscillations of decreasing amplitude. If the ball is pushed hard toward K (analogous to a high birthrate), it overshoots and rolls uphill beyond, perhaps to, point F, whereupon it returns with sufficient velocity to roll not only past K but up and over E and down to 0. K is a stable equilibrium; population density oscillates around K with a tendency to increase when less than K and to decrease when greater than K. E is an unstable equilibrium from which N moves away. However, if $N_0 = E$ at the outset, Equation 4-8 predicts that the population will remain stable; $N_{t+1} = N_t = E$ regardless of birthrate, death rate, or K. Analogously, a ball placed precisely (and precariously) upon point E in Figure 4-3C might remain balanced in unstable equilibrium (until the next earthquake).

STOCHASTIC MODELS

The models developed above, like many theoretical population models and applied models in pest management, are deterministic; that is, they assume a single value for each input, and derive a single value for N_{t+1} after each time interval. A more realistic characterization of actual population events assigns a probability to each demographic measurement. Birthrate, death rate, carrying capacity and extinction threshold all might be defined by probability functions that describe fluctuations about a mean value. Thus, for example, Equation 4-4 becomes

$$N_{t+1} - N_t = N_t \left[P(b) - P(d) \right] \tag{4-9}$$

where $P(b)$ and $P(d)$ are probability functions describing birthrate and death rate, respectively.

In a simplified and extreme example, $P(b)$ and $P(d)$ might be permitted to vary randomly. (Random numbers are assigned to successive calculations of N_{t+1}.) Figure 4-4 shows that simulations using such a model produce an array of nearly meaningless results with but one common element: all become extinct, sooner or later. Mathematically, this must be so because the upper limit is undefined, whereas the lower limit to population density is zero. Biologically, the demography of some local, ephemeral populations sometimes approximates this (Figure 4-3B). From a practical standpoint, modelling of such populations to predict pest outbreaks is futile because they are inherently unpredictable.

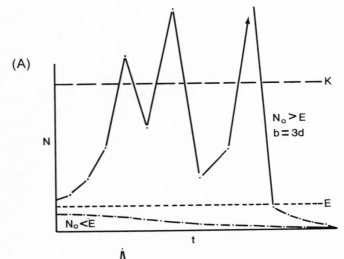

(A)

N

K

$N_o > E$
$b = 3d$

E

$N_o < E$

t

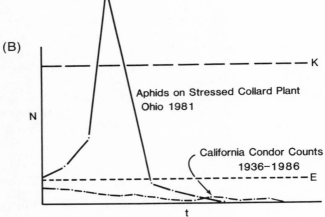

(B)

N

K

Aphids on Stressed Collard Plant
Ohio 1981

California Condor Counts
1936–1986

E

t

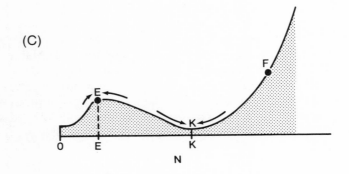

(C)

F

E

E   K   K

0   E   K

N

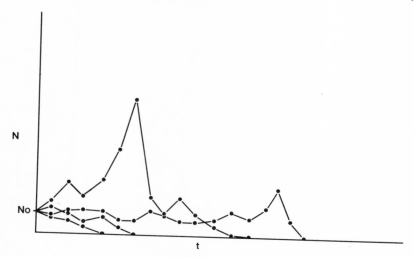

FIGURE 4-4

Results of simulations based on Equation 4-9, exponential model with random birthrates and death rates. Compare tallest peak with aphid data on Figure 4-3B. (Simulations run on NCR Decisionmate V.)

Models using probabilistic functions to define variables are said to be stochastic. They are usually less tractable mathematically than are deterministic models, though in many cases stochastic models supply greater realism in describing actual population events and are thus of some utility in insect pest management (Chapter 12). The widespread use of digital computers has removed the most formidable objection to application of stochastic models of population processes, because the calculation of potential outcomes is no longer a major exercise in tedium (though running complex stochastic simulations can be expensive).

FIGURE 4-3

Logistic model with extinction threshold, E, added. (A) Simulation based on Equation 4-8 with high values of r (rate of increase). (Simulation run on NCR Decisionmate V.) (B) Census data for aphids on stressed collard plant (Columbus, Ohio, 1981) and for California condor population (U.S. Fish and Wildlife Service and State of California). (C) Analogy for modified logistic model; a ball placed right on point E rolls toward K, where it settles after a series of oscillations. A ball pushed hard toward K (high birthrate) overshoots and rolls uphill beyond to point F, whereupon it returns with sufficient velocity to roll past K and over E down to 0.

POPULATION STRUCTURE

Life Tables

The models introduced to this point all assumed that populations contained identical individuals; the term N gave no indication of differences in age, reproductive capacity, genetic makeup, or any of the other characteristics that contribute to the heterogeneity within biological populations. At the outset of this chapter it was noted that individuals within a population are not identical. They dynamics of populations may differ vastly, depending upon a number of characteristics, and age structure is one of the most influential of these. For instance, immature insects often eat more than do adults of the same species, so that the EIL may be lower for immature insects than for adults. On the other hand, adult insects reproduce (by definition), and therefore the birthrate of a population is a direct function of the number of adult females. Immature and adult insects may inhabit entirely different environments, in which they may feed on totally different foods.

It is therefore helpful to subdivide a population into age classes and to analyze mortality separately for each age class. This is accomplished by intensive sampling and expressing the results in the form of a life table, a tabular summary of intergenerational survival and mortality. (In a way, a life table is really a death table.) Table 4-1 illustrates a life table typical of those used by insurance companies to calculate life expectancy for humans and thereby to determine premium rates. Insurance concerns have used human life tables for over 2,000 years, though their use in ecological studies is rather recent, commencing with Deevey's (1947) review. Technically, a life table refers to the fate of a cohort of 1,000 individuals. Richards (1961) proposed "life budget" as a more appropriate term for use in entomological studies based on actual field censuses rather than a uniform 1,000. Ives (1964) and Harcourt (1969) reviewed the earlier applications of life tables to studies of insect populations. A typical life table consists of columns for age class (x), survivorship (= number of survivors entering an age interval, l_x), mortality within an age interval (d_x), and mortality rate within an age interval $(q_x = d_x/l_x)$. Any of these last three columns can be calculated from the other two. Life tables for human populations usually contain an additional column for life expectancy, calculated from survivorship.

A typical insect life table omits life expectancy and gives greater weight to individual mortality factors. Table 4-2 shows a hypothetical insect life table, based loosely on my studies of the alfalfa weevil in

TABLE 4-1
Traditional Human Life Table (Such as Used by Insurance Companies)

Age class (decade)	Survivorship (number alive entering age class)	Mortality (number dying during age interval)	Rate of mortality within age class	Life expectancy (mean lifetime remaining, in decades)
x	l_x	d_x	q_x	e_x
1	1,000	15	0.015	5.75
2	985	25	0.025	4.84
3	960	50	0.052	3.96
4	910	85	0.093	3.15
5	825	125	0.152	2.43
6	700	220	0.314	1.77
7	480	220	0.458	1.36
8	260	150	0.577	1.08
9	110	75	0.682	0.87
10	35	25	0.714	0.83
11	10	8	0.800	0.7
12	2	2	1.000	0.5

Ohio. Age intervals (x) correspond to easily delineated life stages and therefore are not equal in calendar time. Survivorship and mortality are subdivided into components $(d_xF$ and $d'_x)$ to separate the impact of different causes of mortality. Insect life tables often list an additional factor, "killing power" (k_x), calculated from the difference between the log (l_x) entering the age class and log (l_{x+1}), the survivorship at the

TABLE 4-2
Life Table for Hypothetical Weevil Population

x	l_x	d_x	d_xF	d'_x	q_x	k_x
Eggs	1,000	500	Weather	500	0.5	
Larvae	500	250	Weather	150	0.5	0.301
			Parasitism	100		
Pupae	250	200	Weather	100	0.8	0.301
			Predation	100		
Prereproductive adults	50	30	Unknown	30	0.6	0.699
						0.398
Reproductives	20	20	Unknown	20	1.0	

outset of the subsequent age interval. These "k-values" are used in analysis of life tables to determine key factors relating to density changes (see discussion of key factor analysis, below).

Sampling for Life Tables

Sampling insect populations in order to construct life tables is subject to all constraints discussed in the previous chapter. The more accurate the sampling technique, the better it is. In life table studies of insect populations it is customary to attempt to reduce error to about 10% of the mean. A life table for a single generation at a single locality might be interesting, but it is of limited utility in describing the population dynamics of the species under study. Moreover, a single life table is valid only for the specific environmental conditions to which the population was exposed during the census. It is best to sample intensively over several generations and in several locations, ideally over the species' entire geographic range. Obviously, life table studies are long-term, heavily financed ventures, and have been undertaken for relatively few insect pest species.

Life tables may be either age-specific ("horizontal") or time-specific ("vertical"). In age-specific life tables, the fate of a cohort (traditionally, 1,000 individuals) is documented from its collective birth to the death of the last survivor, and survivorship is determined from direct observation. This information is often difficult to obtain for insect populations, because egg, larval (or nymphal), pupal, and adult populations may be sampled using several different techniques with varying degrees of accuracy. Age-specific life tables are statistically valid—that is, the cohort actually represents individuals from a single population and a single generation. In the field, it may be possible to sample the fate of a cohort in those rare instances when oviposition occurs over a short interval and the cohort develops through immature stages more or less in unison. Many insect life tables are age-specific, because most insect populations in temperate climates are seasonal. Age-specific life tables are less feasible when generations overlap, or where there is substantial developmental delay, so that age classes overlap.

Time-specific life tables are concocted from direct samples of the age classes of an entire population, and survivorship is inferred from shrinkage between successive age classes. If a population exhibits a stable age distribution, one would expect more eggs than larvae, more larvae than pupae, and more pupae than adults. The entire population is sampled (usually several times during a season), and the life table is inferred from the proportion alive within each age class. A time-specific life table may be constructed from directly estimated mortality (as that due to predators or parasitoids), and the survivorship is inferred from this. The assumption of a stable age distribution is invalid for highly seasonal species, such as many pest insects (Taylor, 1979), though populations of insects with multiple generations may approach a stable egg distribution once developmental stages overlap. Generally, insect populations in seasonal environments are reset to a single developmental stage annually at the outset of the favorable season, and reach a stable age distribution only gradually, if at all. Thus, a time-specific life table may be valid only late in a growing season, when pests are well-established, whereas pest managers may be most interested in the pest's population dynamics earlier, before the EIL is reached. Southwood (1978) argued that more effort should be spent on developing time-specific life tables for long-lived adult insects (such as mosquitoes) whose fecundity may be an important factor in demography. More practical work needs to be done on determining the age of adult insects.

Most published insect life tables have relied on a combination of both horizontal and vertical data, with most intensive effort having been directed at horizontal life tables.

Construction of Insect Life Tables

EXAMPLE: COLORADO POTATO BEETLE

The Colorado potato beetle (*Leptinotarsa decimlineata*) is a univoltine insect that overwinters as an adult beneath the soil surface. Beetles emerge simultaneously with the appearance of potato leaves after spring planting. Oviposition commences immediately and extends over 3 to 4 weeks. Larvae feed on potato leaves and pass quickly through four instars, then pupate in a cell buried a few centimeters into the soil adjacent to the plants. Adults emerge and feed on the leaves for several days, then enter a summer diapause. Thus, eggs, larvae, and adults are closely associated with observable above-ground portions of the plant, and are sampled rather easily. Moreover, there is a single well-synchronized annual generation, rendering it possible to follow the fate of a cohort.

Harcourt (1963, 1964, 1971) sampled populations of the Colorado potato beetle for 6 years in 10 Ontario locations, and produced a series of life tables, one of which is reproduced, slightly modified, as Table 4-3. Census results were divided into six age intervals, and each hill of

TABLE 4-3

Life Table for Colorado Potato Beetle, Merivale, Ontario, 1963–1964[a, b]

x	l_x	d_x	d_xF	d'_x	$100q_x$	k_x
Eggs	5,643	2,574			45.6	0.265
Larvae, period 1	3,069	1,113	Rainfall	1,113	36.3	0.196
Larvae, period 2	1,956	370	Starvation	370	18.9	0.091
Pupal cells	1,586	58	D. doryphorae	58	3.7	0.016
Summer adults	1,528	1,510	Emigration	1,528	98.9	1.929
Hibernating adults	18	6	Frost	6	33.3	0.176
Spring adults	12	12	Age	12	100	—
						$K = 2.673$

[a]Numbers per 96 hills.

[b]Modified from Harcourt (1971). © Entomological Society of Canada. Used with permission.

potatoes (50 centimeters apart in a row) was a sampling unit. Densities of eggs and larvae were counted per random stalk from each hill sampled. Adults were counted over an entire hill. Densities of pupae were estimated from quadrats of soil dug from each hill. Sampling on each date continued until the confidence limit for the estimate of population density reached 10%. This usually required sampling 200 hills for eggs, 100 to 200 for larvae, 50 for pupae, and 150 for adults.

In constructing life tables, egg density was estimated by direct counting, then inflated by addition of eggs "not deposited" to arrive at an expected number of eggs that would have been deposited had each overwintering female produced her full complement of eggs. This estimate became the size of the initial cohort. Initial larval density was estimated from mortality of eggs. Period-2 larvae were sampled directly, and reduction due to rainfall, a known mortality factor for smaller larvae, was estimated by subtraction. Due to both the extended period of oviposition and the variance in developmental time, larvae were present over several sampling dates. The estimates for larval density represented integration of census data over several sampling dates, a common practice in constructing insect life tables (Figure 4-5; see also Southwood, 1978).

Pupal densities and those of summer adults were based upon direct samples. The tachinid parasitoid *Doryphorophaga doryphorae* was the only major pupal mortality factor, and its readily identified puparia were counted along with pupae of the beetle. Adult beetles

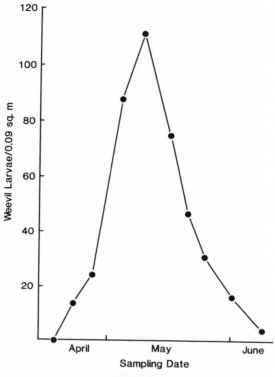

FIGURE 4-5

Estimation of density from area under sampling curve. Alfalfa weevil larvae spend mean of 60 degree-days (10°C base) in second instar. Sampling dates were 80 degree-days apart. Area under curve = 3,200 "larval degree-days"; divided by 60 (developmental time of larvae) = 53.3 larvae/0.09 square meter estimated.

returning in autumn after migration were counted to estimate those hibernating.

EXAMPLE: DOUGLAS-FIR TUSSOCK MOTH

The Douglas-fir tussock moth (*Orgyia pseudotsugata*) is a cyclic defoliator of coniferous trees, principally true firs and Douglas-fir, in the western United States and Canada. Like the Colorado potato beetle, the Douglas-fir tussock moth is univoltine, giving rise to a relatively well-synchronized larval population. However, the Douglas-fir tussock moth is not as easily sampled as is the Colorado potato beetle. Mason (1976, 1981) constructed life tables for declining populations of the Douglas-fir tussock moth in each of 22 locations. Table 4-4 is a compilation from six plots with approximately equivalent levels of defoliation. At each location, biweekly samples of larvae were obtained from each of 15 trees. The average standard error was 22.9%, higher than the ideal of 10%, though the error declined to 10% at high densities, the accurate determination of which was more critical (from the standpoint of insect management).

Instar-I larvae, rather than eggs, were considered the starting point for the cohort. Mason found that eggs were nearly impossible to census accurately. The estimate of egg density (Table 4-4, bottom) was based on estimated fecundity of the adult females present. Samples of larvae were reared in the laboratory to determine the extent of parasitism and disease. Dispersal and predation were inferred from the shrinkage to the subsequent age class. Survivorship of pupae and moths was estimated from mortality within the final larval and pupal stages, respectively.

EXAMPLE: ALFALFA WEEVIL

The alfalfa weevil (*Hypera postica*) is a serious pest of alfalfa throughout most of the United States and southern Canada. In the northeastern United States, there is a single annual generation with a fairly well-defined oviposition period commencing in April. Eggs are deposited in alfalfa stems, and larvae move from there up the stem to feed within the growing plant tip, then move outward onto older foliage. In a heavy infestation they may defoliate an entire field. Pupation occurs in a cocoon on stems or in litter at ground level. Adults usually migrate from the fields in June and spend the summer in wood lots and other hideaways well removed from alfalfa fields.

Harcourt *et al.* (1977) constructed a series of life tables for the alfalfa weevil in Ontario. Oviposition was estimated from the number of punctures in 64 to 128 three-stem samples taken twice weekly. Viability and mortality were estimated directly from egg samples cut from these stems. Larvae were extracted via berlese funnels from 64 six-stem samples taken twice weekly. Mortality was measured directly from rearing of parasitoids until epizootics of the fungus *Erynia* sp. (*Entomophthora phytonomi* in the 1977 paper) rendered it more practical to estimate mortality from shrinkage between age classes. Estimates for prepupae and pupae were made from direct counts of cocoons in the foliage. Table 4-5 shows the resulting composite life table.

TABLE 4-4

Life Table for Doublas-Fir Tussock Moth on Light Defoliation Plots, 1973–1974[a]

x	l_x	d_x	d_xF	d'_x	$100q_x$
Instar I	479.7	250.0	Predation, dispersal	250.0	52.1
Instar II	229.7	170.7	Virus	3.0	74.3
			Parasitoids	0.9	
			Predation, dispersal	166.8	
Instars III, IV	59.0	38.4	Virus	1.1	65.1
			Parasitoids	1.1	
			Predation, starvation	36.2	
Instars IV, V	20.6	8.8	Virus	1.9	42.7
			Parasitoids	1.6	
			Other	5.3	
Instars V, VI	11.8	2.1	Virus	1.9	17.8
			Parasitoids	0.2	
Pupae	9.7	7.0	Virus	1.1	72.2
			Parasitoids	4.7	
			Other	1.2	
Moths	2.7		Sex ratio (35% females)		
Females × 2	1.89	0.5	Reduced fecundity	0.5	26.2
Normal females × 2	1.39	1.39	Aging	1.39	100
Eggs	103.5	99.7	Parasitism	50.8	96.3
			Infertility	17.0	
			Stress, predation	31.9	
Expected instar I	3.9				

[a]Mason (1976). © Entomological Society of America. Used with permission.

In Ohio, Lewis (1977) constructed alfalfa weevil life tables by a somewhat different procedure. No estimation of egg density gave consistent accuracy, so that egg density was inferred from counts of period-I larvae, to which was added loss of eggs due to infertility and parasitism. Larval and pupal densities were estimated from ten 0.09-square-meter weekly samples of foliage, and ground litter was extracted by berlese funnel. Neither Lewis nor Harcourt *et al.* were able

TABLE 4-5

Within-Generation Life Table for Alfalfa Weevil: Mean Values for 15 Plot-Years, Quite Area, Eastern Ontario, 1972–1976[a, b]

x	l_x	d_x	d_xF	d'_x	$100q_x$
Eggs	421	9	*Patasson luna*	1	2.1
			Infertility	8	
Period 1	412	106	Establishment	106	25.7
Period 2	306	261	*E. phytonomi*	255	85.3
			Rainfall	6	
Prepupae	45	10	*E. phytonomi*	7	22.2
			Bathyplectes curculionis	2	
			Tetrastichus incertus	1	
Pupae	35	4	*E. phytonomi*	4	11.4
Summer adults	31	30	Winter disappearance	30	96.8
Spring adults for $R_0 = 1$	1				

[a]Numbers per 0.9 square meter.

[b]Harcourt *et al.* (1977). © Entomological Society of Canada. Used with permission.

to adequately census adult population densities, so these were inferred from estimates of pupal mortality.

Table 4-6 is a composite life table for Ohio alfalfa weevils. Despite the differences in sampling technique, there is correspondence between the Ohio and the Ontario life tables. The fungus *Erynia*, found by Harcourt *et al.* to be a key factor influencing population trends in Ontario, was undetected in Ohio, nor was the unimportant parasitoid *Tetrastichus incertus*. In Ohio, the parasitoids *Bathyplectes anurus* and two species of *Microctonus* were well-established and contributed minor mortality. These differences underscore the need for intensive population studies throughout the entire geographic range of an insect pest before overall generalizations can be made legitimately about its population dynamics.

UTILITY OF LIFE TABLES

Calculation of Replacement Rate

A valid life table can be used to determine whether a population is growing, declining, or remaining stable. The replacement rate, R_0, is calculated from survivorship of adult females and the average fecundity of females (m_x, fecundity per individual female). $R_0 = \Sigma l_x \cdot m_x/$ 1,000, and equals the rate at which each egg in the cohort at the outset is replaced by an egg entering the subsequent generation. In Table 4-2, for example, I assumed that the sex ratio was 50:50, so that there were 10 females at the outset of reproduction, and that each female produced 100 eggs. In this case, $R_0 = 10 \times 100/1,000 = 1$ and the population size is constant, neither increasing nor declining.

Simulation

Once a valid life table is constructed for an insect population, it may be used to simulate the outcome of management decisions whose effects may be more complex than is apparent at first. For example, suppose that we wish to manage the hypothetical weevils of Table 4-2 with an insecticide application that increases mortality of adults but (perhaps unfortunately) eliminates all parasitoids and predators. (This is oversimplified, yet some real-life management schemes produce very similar effects.) We further assume that the effect of weather is density-independent (removing the same proportion of the population regardless of its density). The question is: Would this population increase, decrease, or remain stable?

TABLE 4-6

Within-Generation Life Table for Alfalfa Weevil: Mean Values for 11 Plot-Years, Central Ohio, 1973–1976[a, b]

x	l_x	d_x	d_xF	d'_x	$100q_x$
Eggs	246	26	*Patasson luna*	10	10.6
			Infertility	16	
Period 1	220	81	Establishment loss	81	36.8
Period 2 and prepupae	139	18	*Bathyplectes curculionis*	14	13.0
			B. aunurus	4	
Pupae	121	7	Unknown	7	5.8
Summer adults	114	113	Winter mortality	109	99.1
			Microctonus spp.	4	
Spring adults for $R_0 = 1$	1				

[a]Numbers per 0.9 square meter.
[b]Lewis (1977).

Intuitively, one might think that the population would decrease; after all, mortality of prereproductive adults is now 90%. Does this added mortality compensate for the loss of predation and parasitism of larvae? Table 4-7 is a modification of Table 4-2, and takes into account the assumptions above. There are now 22 prereproductive adults from every 1,000 eggs, and if 50% were females, averaging 100 eggs each, R_0 = 1.1, and the population would increase in every generation.

Calculation of r_m and r_c

In developing models of population dynamics, one may wish to know the maximum rate at which a population is capable of increasing in a favorable environment. This is an especially crucial parameter in temperature-dependent models, wherein we wish to know the speed at which the EIL is reached at various temperatures.

For populations with a stable age distribution, r_m may be calculated from the life table by solving

$$\sum e^{-r_m x} (l_x) (m_x) = 1 \qquad (4\text{-}10)$$

for all age classes in which reproduction occurs. A series of life tables constructed from laboratory populations reared under several different temperatures can be used to determine the optimum temperature for maximum population growth. Messenger (1964) and Herbert (1981) used this method to calculate r_m for spotted alfalfa aphids and European red mites, respectively.

When a stable age distribution cannot be assumed, especially among seasonal insects, r_m may be approximated by the capacity for increase, r_c

$$r_c = \log_e R_0 / T \qquad (4\text{-}11)$$

where T is mean generation time, from egg to mean adult longevity. For the hypothetical weevils of Table 4-7, $r_c = \log_e 1.1/1 = 9.5\%$ annual increase. (This is lower than the actual r_c for many insect pests.)

The statistics r_m and r_c are characteristic for each insect species under the appropriate environmental conditions and are useful indicators of the speed with which a pest may be expected to reach the EIL. Cave and Gutierrez (1983) illustrated the utility of this concept by calculating r_m for *Lygus* bugs on alfalfa and cotton. The r_m was higher on alfalfa, indicating it was a more suitable host and might be useful as a "trap crop" to draw *Lygus* away from cotton (Chapter 12).

TABLE 4-7
Modification of Life Table (Table 4-2) for Management Simulation

x	l_x	d_x	d_xF	d'_x	q_x
Eggs	1,000	500	Weather	500	0.5
Larvae	500	150	Weather	150	0.3
Pupae	350	140	Weather	140	0.4
Prereproductive adults	210	188	Insecticide + unknown	188	0.9
Reproductives	22	22	Unknown	22	1.0

Determination of Key Factors

Introduced by Morris (1959, 1963) key factor analysis has proven to be a valuable aid in identifying the environmental factors most closely related to intergenerational population trends. Key factor analysis begins with a series of life tables extending over several generations of a population, usually from a single location. The life tables already discussed for Colorado potato beetle (Table 4-3), Douglas-fir tussock moth (Table 4-4), and alfalfa weevil (Tables 4-5 and 4-6) are all suitable for key factor analysis.

Initially, Morris (1959) used the relationship between each l_x and total survivorship (calculated from the product of all the l_x's). Varley and Gradwell's (1960, 1965, 1968) graphical key factor analysis is today generally considered superior (Luck, 1971; Southwood, 1978). They used k_x values summed over each generation to arrive at a total K. Life tables from adult-to-adult are preferred for this analysis. Each k_x is plotted over several generations, along with the total K, and the relative importance of each key factor is evident from its similarity to the plot of total K. Table 4-8 and Figure 4-6 illustrate this for Harcourt's life tables of the Colorado potato beetle from 1961 to 1966. The curve for k_5 (adult emigration) bears the greatest similarity to the overall mortality curve; therefore, the key factor should be sought in the fifth age interval. Larval starvation (k_3) is more weakly related to total K.

Podoler and Rogers (1975) offered a modification of the graphical method of Figure 4-6. In their method, common K is plotted against each k_x. The closer the apparent relationship, the greater is the contribution of the k_x to total K. Figure 4-7 shows Harcourt's data analyzed by the technique of Podoler and Rogers. Again it is evident that a key factor is influential within the interval when emigration occurs (k_5).

There are three cautions in using this analysis.

TABLE 4-8
k-Values for Harcourt's (1971) Life Tables of Colorado Potato Beetle

Stage	1961	1962	1963	1964	1965	1966
k_1 (eggs)	0.234	0.222	0.265	0.303	0.235	0.364
k_2 (larvae 1)	0	0.094	0.196	0.074	0.099	0.008
k_3 (larvae 2)	0.337	0.149	0.091	0.099	0.191	0
k_4 (pupal cells)	0.002	0.011	0.016	0.023	0.013	0.058
k_5 (summer adults)	2.295	2.204	1.929	1.429	3.038	0
k_6 (hibernating adults)	0.058	0.135	0.176	0.521	0.135	0.075
K	2.926	2.815	2.673	2.449	3.711	0.505

1. Individual k's are included in the calculation of total K, so that statistical correlation and regression are invalid. Total K is a dependent variable by definition, and statistical correlation is valid only for independent variables.

2. A relationship between total K and any k_x is only an indicator. Key factors are useful predictors of population trends, but may not be the actual causes of the observed trends. Additional intensive study on the action of various environmental factors must be undertaken in order to prove causation.

3. A key factor operating in one location may not be a key factor in another location. Recall that Harcourt et al. (1977) found the fungus *Erynia* sp. to be a key factor in population trends of the alfalfa weevil in Ontario, yet in Ohio Lewis (1977) never encountered the fungus. In a study of the winter moth for over 18 years in England, Varley and Gradwell (1968) found that winter disappearance was a key factor and that mortality due to the parasitic fly *Cyzenis albicans* was insignificant, yet after the winter moth was accidentally introduced into Nova Scotia, *C. albicans* was also introduced and within a few years became the major mortality factor reducing populations of the winter moth in eastern North American (Embree, 1966, 1979).

The effect of some environmental factors (particularly predation, disease, and competition) may be largely density-dependent; that is, their influence on the population varies in proportion to the numbers present and is generally more severe in crowded populations. Other factors (weather, insecticides, soil tillage, burning of crop residues, etc.) exert primarily density-independent effect, and, though capable of reducing numbers, generally are incapable of regulating density about an equilibrium. Laboratory populations generally exhibit more evident density-related mortality than do populations in the field.

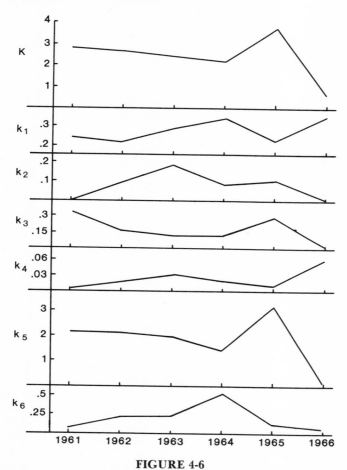

FIGURE 4-6

Key factor analysis for Harcourt's (1971) Colorado potato beetle life tables (Table 4-8), analyzed by method of Varley and Gradwell (1960). The plot of k_5 (emigration) versus time shows greatest similarity to that of K versus time (topmost line). Therefore k_5 is the most likely key factor.

Strong (1984) suggested that regulation of field populations is "density vague" due to high variance and time lags. All populations are influenced by a complex of factors varying in their impact on density. However, key factor analysis can reveal the extent of density dependence. Each k_x is plotted against survivorship (l_x) at the start of the age interval. Positive slope indicates density dependence; the closer the slope is to 1, the more likely it is that the factor is capable of stabilizing

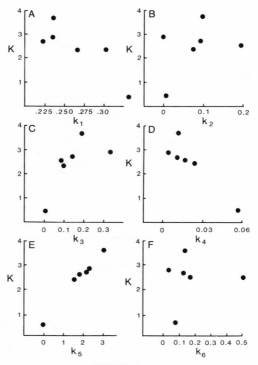

FIGURE 4-7
Key factor analysis for Harcourt's (1971) Colorado potato beetle life tables (Table 4-8), analyzed by method of Podoler and Rogers. (A) Egg mortality, k_1 (weak relationship with K). (B) Rainfall, k_2 (no relationship with K). (C) Starvation, k_3 (weak relationship with K). (D) Parasitism by *Doryphorophaga doryphorae*, k_4 (negative relationship with K). (E) Emigration, k_5 (close relationship with trend of K indicates key factor). (F) Frost, k_6 (no relationship with K). (See text for further explanation.)

the population. Slopes greater than 1 indicate overcompensating density dependence, whereas slopes less than 1 indicate undercompensating density dependence. A negative slope is an indication of inverse density dependence. Density-independent mortality is characterized by lack of any apparent relationship between k_x and l_x. Figure 4-8 shows these density relationships for Harcourt's Colorado potato beetle study. From this analysis one can see that starvation (k_2) and emigration (k_5) are positively density-dependent, that pupal parasitism (k_4) is negatively density-dependent, and that egg mortality and mortalities due to rainfall and frost are density-independent. Strong *et al.* pointed out that this methodology is imprecise; key factor analysis may indicate density dependence when none is present, or miss weak density dependence when it is there.

Royama (1981a, 1981b) and Southwood (1978) give further details and methodology on separation of density-independent from density-dependent factors in life table analysis. Studies such as those of Danthanarayana (1983) on the light-brown apple moth, of R.B. Ryan (1983) on the larch casebearer, and of Wright *et al.* (1984) on bark

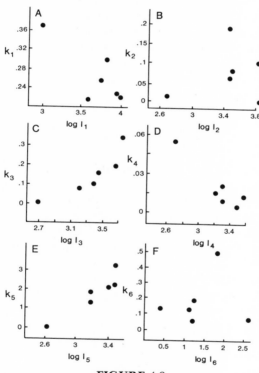

FIGURE 4-8

Harcourt's (1971) Colorado potato beetle life table data analyzed for density dependence. (A) Eggs, no density dependence. (B) Larvae, period 1 (rainfall), no density dependence. (C) Larvae, period 2 (starvation), strong density dependence. (D) Pupae (parasitism by *Doryphorophaga doryphorae*), negative density dependence. (E) Summer adults (emigration), strong density dependence. (F) Hibernating adults (frost), no density dependence.

beetles attest to the value of key factor analysis in elucidating the relative impact of various environmental factors on pest population dynamics.

Limitations

As has been said, life table analysis is only as valid as the accuracy of the sampling techniques used to obtain the initial data. One must regularly reascertain the validity of the assumptions underlying sampling techniques themselves as well as resulting statistical calculations. Especially critical are the relationships among the various mortality factors. For simplicity, mortality is usually considered constant within each age class, and mortality factors are considered sequential and mutually exclusive. For instance, Lewis's alfalfa weevil life table (Table 4-6) assumed that parasitism by *Bathyplectes curculionis* preceded that by *B. anurus*, and that multiple parasitism did not occur. (It did, though not often, and the activity periods of the two species do overlap in the field.)

GENETIC SYSTEMS

Genetic variation plays an important role in population dynamics of insect pests and in designing effective strategies for ecological pest management. With the grand exception of the pomace fly, *Drosophila melanogaster* (and other *Drosophila* species), whose genome has been more thoroughly scrutinized than that of any other animal (humans included), the details of genetic systems are not well understood even for common insect species, including most pests. The genetics of the commercial silkworm and the honeybee are known in some detail, however, and there are ambitious predictions of directing changes in insect genomes in order to manage pests more efficiently. Analysis of insect genetics has become a valuable tool as a means of delineating the limits of local populations within a species. For example, Pashley *et al.* (1985) analyzed proteins in *Spodoptera frugiperda* electrophoretically, and concluded that populations in the southeastern United States were the offspring of moths overwintering in Mexico rather than in the Caribbean, as had been previously assumed. Pashley and Bush (1979) determined through allozyme studies that east-central Europe was the ancestral home of the codling moth because of its high genetic variation there.

Like all organisms, terrestrial arthropods display behavioral, physiological, and morphological traits in the phenotype that are under the control of an underlying genotype. Almost all insects are diploid (important exceptions are haploid males in Hymenoptera and some Homoptera and Acari), generally displaying dominance and recessiveness at each genetic locus according to the expectations of classical genetics. Every population exhibits genetic heterogeneity, and this may be reflected in a high variance when enzymes are analyzed (Lewontin, 1974). This polymorphism seems to increase with environmental heterogeneity (Nevo, 1978). Many agricultural ecosystems are notoriously variable, perhaps selecting for high genetic variance on the part of their insect (and weed) inhabitants. High fecundity and short life cycles are typical of many agricultural pests (so-called "r-strategists"; Chapter 5), enhancing tendencies toward polymorphism by increasing potential recombination. Genetic variance also changes as populations cycle, and is usually greatest in high-density populations (Alstad & Edmunds, 1983).

Selection

An elementary principle of biological science and one of the cornerstones of the modern evolutionary synthesis is that natural selection favors genes conferring positive adaptive traits, and that this process results in morphological, physiological, and biochemical changes evident in the phenotype. The case of insecticide resistance is an outstanding example of the process (Chapter 7), though in many other instances it is difficult to ascertain precisely the selective pressures, just as it is hard to isolate a single key factor in population dynamics. Most often, populations respond genetically to a complex of environmental pressures, resulting in evolution of organisms representing a compromise rather than a single, obviously "fittest" type.

The impact of selection may be much more subtle than that induced through a factor such as insecticide resistance. Planting of resistant varieties of crops (Chapter 10) may select in favor of genetic "biotypes" among insect pests, whereupon the pest appears to have overcome the resistance. "Biotype" in its narrowest sense implies a population that has an identifiable genetic change corresponding to a gene for resistance in a host plant (Claridge & den Hollander, 1983; also Chapter 10). An example is that of the greenbug, *Shizaphis graminum*, of which several biotypes exist. Planting of resistant wheat varieties has led to selection favoring specific biotypes, and plant resistance has sometimes become ineffective as a management technique (Gallun *et al.*, 1975). A similar predicament may have arisen in the case of a screwworm (Chapter 11), in which releases of sterile males from laboratory cultures have apparently resulted in selection favoring specific local genetic races in Mexico.

Pimentel (1961a) proposed that predator–prey and host–parasite systems are sufficiently coevolved that genetic changes in each might result in evolution of prey (or host) tolerance to a natural enemy, and the predator or parasite in turn might become less virulent toward its prey. This is certainly possible in microbial control of insects (Chapter 8) and may occur with "classical" biological control as well (Chapter 9). For example, Muldrew (1953) found that the pine sawfly evolved the ability to encapsulate eggs of its parasitoid *Mesoleius tenthredinis* after 20 generations, resulting in less effective biological control.

Most elementary models of population genetics assume random mating. For mathematical tractability it is convenient to assume that each individual has an equal chance of mating with any other individual (of opposite sex, of course) in the population. For many insect populations this assumption is invalid, and the greater the amount of local inbreeding, the greater is the likelihood of locally selected races' arising. This is even more likely if dispersal is reduced at the same time. Many parasitic Hymenoptera mate immediately upon emergence of the female, which greatly increases the incidence of inbreeding; it may well be that such a mating system retards the development of insecticide resistance among agents of biological control (Horn & Wadleigh, 1987). In insects (such as Hymenoptera) with haplodiploid sex determination, recessive genes are expressed in males of each generation, enhancing the spread of these genes in the population (Georghiou & Taylor, 1977).

RESPONSES TO ENVIRONMENT

Dormancy

Insects and their relatives are small and poikilothermic and therefore very much at the mercy of the physical environment. In areas of seasonal cold and/or drought (which include most of the earth), adaptations that permit insects to avoid maladaptive conditions have evolved. Seasonal dormancy (hibernation, if in winter; aestivation, if in summer) can occur in any life stage, though in most cases it is the egg or pupa that passes the inclement season. In simple quiescence, body processes slow in direct response to lowered temperature or lack of moisture. Examples are some mosquitoes, blowflies (Calliphoridae), and the housefly. In the midwestern United States, any of these can be stirred into action on mild winter days; sunshine and an ambient temperature of 16°C brings *Calliphora* blowflies to my porch in Columbus, Ohio, even in January.

Insects may also enter diapause, a physiological resting state wherein there is an obligate, hormonally mediated slowing of physiological development and growth, in response to an environmental cue. In most cases it is diminishing photoperiod that triggers hormonal changes (usually a reduction in juvenile hormone production) resulting in cessation of activity before the onset of harsh conditions. Denlinger (1981) and others have shown that induction of diapause is related to juvenile hormone activity in flesh flies and hornworms. In some insect pest species, diapause is fixed and obligate in every generation. Such insects are termed "univoltine." Acridid grasshoppers of the temperate zone undergo diapause in the egg stage, as does the gypsy moth. The alfalfa weevil and Colorado potato beetle each have an obligate adult diapause. "Multivoltine" insects are those that have two or more annual generations and that enter diapause only at specific times. The European corn borer larva enters diapause for the overwintering generation but not for the midsummer generation. In some adult insects, diapause does not result in cessation of activity, but, rather, causes delayed development of the gonads while stimulating increased migratory behavior (see below). An intriguing possibility for insect management is disruption of normal diapause to expose pests to inclement conditions; this has yet to be accomplished in the field.

Dispersal and Migration

Dispersal (sometimes as a migratory diapause) is an advantageous adaptation to countering the ephemeral availability of resources facing most insect populations. Dispersal may be directed movement to seasonally available resources (in which case it is truly migration) or largely random movement outward from centers of high density. Dispersal enhances (1) escape from crowded conditions, (2) colonization of uninhabited but suitable areas or food plants, and (3) increased outbreeding and genetic recombination, resulting in greater evolutionary plasticity. Dispersal places insects at risk, from predation, inclement weather, or simply getting lost. Johnson (1969) remains a classic comprehensive study of insect dispersal, though an updated treatment is due.

Generally, insect dispersal increases under crowded conditions or in response to declining quality and quantity of resources. Many pest species are polymorphic, containing both dispersing and sedentary forms (Harrison, 1980). Denno (1977, 1979, 1983) studied the interactions among habitat stability, population density, and dispersal in the saltmarsh planthopper *Prokelisa marginata*. The insect feeds on sap from salt-marsh grasses (*Spartina* spp.) abundant along the Atlantic and Gulf Coasts of the United States. The planthopper exhibits two forms: a short-winged morph (with higher fecundity) and a longer-winged, dispersing morph. The frequency of each is directly related to seasonal stability of habitat; crowding plus reduced host nutrition results in a higher population of longer-winged forms.

A classic case illustrating the role and control of dispersal in insect population dynamics is that of the migratory locust, *Schistocerca gregaria*. Uvarov (1966), White (1976), and Rainey (1982) discuss locust dispersal in detail. The locust is widespread in the semiarid areas of the Old World tropics and subtropics, from whence swarms sometimes spread as far as the Mediterranean region. In the nonmigratory phase (*solitaria*), locusts are wingless as adults, and movement is strictly local. As the food supply (mostly grasses) improves, the locusts become more abundant and, under the influence of crowding, juvenile hormone production is reduced and adults develop functional wings and flight muscles. (This is inducible experimentally by agitating locusts during nymphal development.) The migratory phase (*gregaria*) takes flight in huge swarms that are carried by wind to frontal convergence zones where rain is likely to be falling, resulting in lush grasses that can then sustain the locust population. This is a very effective adaptation for assuring survival in a semiarid climate wherein rainfall and food supply are unpredictable, though it is not appreciated by those who are attempting to grow grain crops in the path of the swarms. Locust problems can be acute in semiarid areas of central and eastern Africa, where a growing human population attempts to produce adequate food under conditions that are marginal at best.

Many agricultural pests of the northern United States and Canada cannot survive winter conditions there. Species such as the potato leafhopper and the corn earworm undertake a seasonal one-way northward migration each spring from overwintering sites in the southern states. An interregional network of sampling via light traps and/or sweep netting can document these movements to predict the onset of damaging populations in northern areas.

Dispersal may be related to qualitative variation in a single population at a single time. Wellington (1979) noted that populations of the western tent caterpillar (*Malacosoma pluviale*) consisted of both active and inactive morphs (apparently a genetic polymorphism). Active forms tended to disperse outward and founded new infestations, while inactive morphs remained behind, whereupon they induced changes in host plant chemistry sufficient to greatly reduce local survivorship. The process was a factor contributing to cyclic changes common in forest insects (Chapter 5).

Dispersal also may occur in seasonal pulses, often in response to variations in availability of host plants. In my studies of aphids, I consistently note a bimodal immigration onto experimental collard plants in response to the decline in quality of nearby *Brassica* spp. as suitable hosts for cabbage aphids.

Dispersal is a most important component of insect population processes, yet, in general, spatial dynamics of insect populations are not as well-studied (or as well-modeled) as are temporal dynamics (Kennedy & Margolies, 1985). As a result, the process often appears to be so variable as to defy incorporation into models useful in predicting insect outbreaks for pest management decisions (Stinner *et al.*, 1983). Movement is extremely important in understanding the dynamics of insect pests, both locally and regionally. Kareiva (1983b) adopted a simple passive diffusion model that adequately explained local dispersal of several species of phytophagous insects, and he found a wide variation in diffusion coefficients. Such variability has important implications for the speed at which EIL is reached locally, the effectiveness of quarantines, and the spread of insecticide resistance. Insect dispersal is a fruitful area for research, and much is to be gained by concentrated interregional and international research efforts relating to the movement of arthropod pests (Taylor *et al.*, 1980).

CHAPTER

5

INSECTS IN
ECOLOGICAL COMMUNITIES

Insect populations are integral parts of natural and anthropogenic (man-made) ecosystems, and any attempt to reduce and maintain insect densities below EILs must consider the role of environmental factors that influence pest numbers. The ultimate failure of many insect control programs based exclusively on chemicals is traceable in part to a cavalier disregard of the surrounding ecological community in which the pest population occurs.

Clark *et al.* (1967) proposed the term "life system" to connote an insect population plus its effective environment. The effective environment includes intrinsic and extrinsic factors; competition, predation, disease, weather, food supply—in short, any and all environmental factors that influence births, deaths, immigration, and emigration, and that thereby influence population density. The effect of some environmental factors may be density-dependent (Chapter 4), whereas other factors exert primarily density-independent effects.

Ideally, a pest manager wishes to manipulate the life system of a pest in such a way as to reduce density by increasing the negative influence of environmental factors, to set a new equilibrium below the EIL (Figure 2-1). Because an entire ecological community must be managed at once, successful pest management must consider interactive models including, where appropriate, weather, plant growth, weeds, pathogens, natural enemies, and potential competitors of the pest.

INTERSPECIFIC INTERACTIONS

Broadly, there are five types of interactions with other organisms, though the categorization exists as much for the convenience of ecologists as for any other reasons. Mutualistic interactions (++) benefit both populations. A trivial example is that between bees and flowers, each of which depends upon the other: the bees gain nutrients and the flowers are pollinated. Competition (−−) is the opposite, a negative effect upon both populations. An example is the interaction between two species of grain-feeding beetles in a flour bin in which food is limited. Exploitation (+−) advances one population at the expense of another. Predation is an obvious example, though competitive interference can also be exploitative. Commensalism (+0) benefits one species with no discernible effect on another, as in the case of the many nearly microscopic arthropods that disperse by attaching themselves to a larger insect, bird, or mammal. Neutral interactions are also possible, though one might argue in such cases that there is no "interaction" at all.

Of these interactions, predation has received the most attention from insect ecologists, perhaps due to its demonstrated utility in pest management. Further study may reveal some intriguing possibilities for manipulating competitive, mutualistic, or commensal associations for more effective pest management.

Predation

In its broadest sense, predation includes any interaction wherein one species benefits at the expense of a second species by obtaining nutrients directly from the second species. This definition includes pathogenesis and parasitism as well as the activities of more traditionally familiar predators, such as lions, tigers, and bears. By this definition, herbivores may be considerd predators of plants (Chapter 6). Because predation is capable of reducing density of at least some pest populations, it has been manipulated for many decades to achieve partial or complete biological control (Chapter 9).

Generalizations about such a ubiquitous phenomenon as predation are hard to come by, though characteristically, predation is a density-dependent process. A simple model describing predation adds a component to Equation 4-6:

$$N_{t+1} = N_t + N_t (b - d) \left(\frac{K - N_t}{K} \right) - cP_t N_t \qquad (5\text{-}1)$$

where P_t is predator population at time t, and c is a constant, equal to the proportion of P_t that actually kills and devours an individual of N_t during the interval t to $t + 1$.

A simple model for the corresponding predator population is

$$P_{t+1} = P_t + cP_tN_t - dP_t \qquad (5\text{-}2)$$

where c is a constant, as in Equation 5-1, and d is the predator's death rate (assumed to be constant, for simplicity). This model carries the simplifying, though usually false, assumption that there is a 1:1 conversion of prey to predator. This occurs only in the specialized case of solitary insect parasitoids, mostly wasps or flies. "Classical" insect predators (e.g., lady beetles and mantids) eat many more than one prey individual to replace themselves, and, to adequately model their demography, the constants in Equations 5-1 and 5-2 must be adjusted accordingly. However, the simplified model does provide a useful insight into the nature of predation, as is evident from Figure 5-1. This model illustrates the density-dependent nature of predation, wherein the predator population, totally dependent upon prey density, oscillates with it in a series of coupled cycles, out of phase with one another due to the time delay before predator reproduction.

The model also assumes that rates of predation per individual predator are constant (c). Occasionally, this may occur, though more often predators or parasitoids demonstrate a variable response to den-

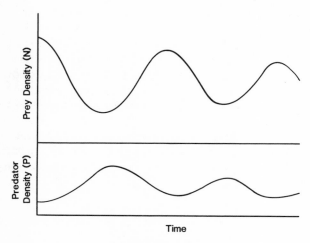

FIGURE 5-1

Coupled predator–prey oscillations from simulation based on Equations 5-1 (prey) and 5-2 (predator). The populations cycle out of phase with one another.

sity of prey or host. Figure 5-2 illustrates this "functional" response (Holling, 1959), the relationship between prey/host density and the rate of attack. Functional responses have been characterized broadly as being of three types, all of which show reduced (or no) predation at very low densities of prey and an upper limit (equal to c of Equation 5-2), due to handling time for each prey item, and eventually due also to satiation. Between these limits, the shape of the curve may differ for various predator species, and this can have great practical influence on the efficiency of biological control. A Type I curve is characteristic of a monophagous parasitoid or predator. There is a threshold level (below which prey are too rare for attack), followed by a steep rise to an inflection point representing satiation; the predator is devouring prey at maximum rate and is limited beyond that point by its stomach capacity, if nothing else. Such a response has been found among many insect parasitoids in laboratory studies (Hassell, 1978). Type II is an intermediate response also characteristic of monophagy. A sigmoid functional response (Type III) is more typical of predators (especially vertebrates) that subsist on a variety of prey species and among which learning plays a role in increasing the efficiency of predation. At low densities of prey, a predator exhibiting a Type III response may switch to a more readily available prey species yet continue to prey infrequently on the rarer species.

Hassell (1978; Hassell et al., 1977) has shown theoretically that only Type III responses stabilize a predator–prey interaction. Models of functional response are destabilized by inclusion of age-structure (Wollkind et al., 1980), yet addition of alternative prey species tends to restabilize the relationship (at least in computer simulations) (Inouye, 1980). Because insects are poikilothermic, ambient temperature is an important factor determining the nature of the functional response. Mack et al. (1981) developed models that included temperature in functional response curves. The role of temperature may be especially critical when the activity threshold for prey is below that for predators; populations of aphids may increase in cool wet weather that slows or halts activity of predatory Coccinellidae and parasitic Hymenoptera.

Many functional response curves have resulted from laboratory studies, whereas functional responses are notoriously difficult to measure in the field (Hassell et al., 1977; Horn & Dowell, 1979). Moreover, the "classical" functional response model of Holling (1959) makes numerous simplifying assumptions, such as that prey are distributed randomly, that prey are not depleted, and that predators' handling times are constant (Houck & Strauss, 1985). Berryman (1967) pointed out that a change in density of prey and time of exposure to predators alters the shape of the functional response curve. Increasing prey

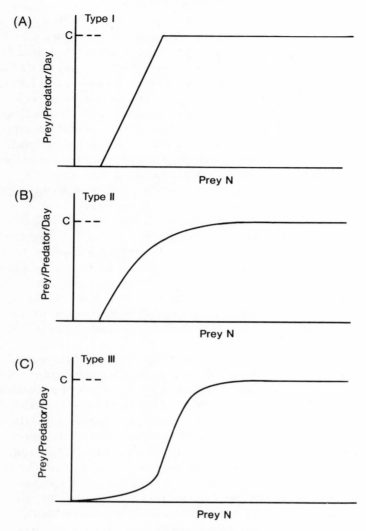

FIGURE 5-2

Three types of functional responses of predators to prey density: Type I, characteristic of monophagous predators; Type II, intermediate; Type III, characteristic of polyphagous vertebrate predators, most likely to stabilize predator–prey interaction. All level off at C, level of satiation. Modified from Holling (1959). © Entomological Society of Canada. Used with permission.

density may reduce the searching efficiency of individual predators, due to their interactions with one another (Podoler & Mendel, 1979). This tends to reduce c in the model (Equation 5-2). Moreover, real predators do not switch as readily to commoner hosts as computer simulations predict. Coccinellidae and *Zetziella* mites continue to eat prey species to which they have grown accustomed and switch to commoner prey only with great reluctance and after some delay (Murdoch & Marks, 1973; Santos, 1976). The parasitoid *Nasonia vitripennis*, a wasp whose larvae feed on pupae of muscoid flies, also exhibits delayed host-switching (Cornell & Pimentel, 1978). Such results suggest that functional responses derived from laboratory studies represent optima but that environmental heterogeneity in the field leads to time delays and to an overall reduction of the functional response. The overall outcome is that fewer prey are eliminated than one would expect. Nevertheless, functional response is a useful concept that provides insight into the complexities of predation.

By analogy with the functional response, the numerical response is the increase in predator or parasite density in response to increase of the prey or host (Figure 5-3). Like the functional response, the numerical response varies in form, but characteristically has a lower bound below which prey is too scarce to support a predator population, and an upper limit where interactions among predators themselves (territoriality, cannibalism, availability of resting or breeding sites) prevent further increase of the predator population, regardless of how much prey is available. In determining the impact of predator upon prey, these lower and upper limits are probably more important than the shape of the curve between.

Functional and numerical response can be combined into a total response (Figure 5-4) modeling the relationship of predator to prey density. Of greatest importance here is that the density-dependent effect of predation may be exerted over only a limited range of prey densities (N_1 to N_2). Below N_1, prey is too scarce to support the predator population, which will die out or emigrate. Above N_2, predators are fully satiated and reproduce at maximum capacity but no longer exert a regulating effect on populations of prey. The prey, at densities above N_2, effectively "escapes" regulation by outbreeding the predator. (This is discussed further below.)

The precise role of predation in limiting pest densities in the field is a matter of debate (Dempster, 1983; Hassell, 1985). Examples of successful biological control (Chapter 9) attest to the occasional utility of predators, and consideration of predator–prey interactions is important in selecting agents of biological control. In particular, parasitoids

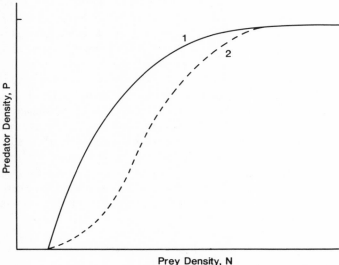

FIGURE 5-3
Numerical response of predator density to prey density; form varies in laboratory studies from curvilinear (1) to sigmoid (2). Sigmoid displays strongest density dependence. Upper limit determined by maximum reproductive capacity of predator.

(insects, usually Hymenoptera or Diptera, with a free-living adult and a parasitic larva) are more specific in their choice of hosts than are more general predators. In terms of the model in Figure 5-4, the "total response" shifts leftward, and density-dependent mortality (N_1 to N_2) occurs at a lower host density. However, the host does not (usually) die or even cease activity until the parasitoid larva is nearly fully grown, so there is a longer delay in any observable impact of parasitism on population density of the host and its damage to the crop. Predators, on the other hand, kill and consume prey at much faster rates than do analogous parasitoids, and are also more likely to survive periods of host scarcity by switching to alternative foods.

Microbial pathogens are a special case, for they have no actively free-living stage searching for hosts or prey (though some pathogens are spread and inoculated on the ovipositors of parasitic wasps, while others induce infected hosts to greater dispersal activity). The effect of microbes is a delayed density-dependent one; in terms of Figure 5-4B, N_1 is shifted to the right. Effects of pathogens may also be overcompensating; that is, as an epidemic spreads, a very large proportion of

(A)

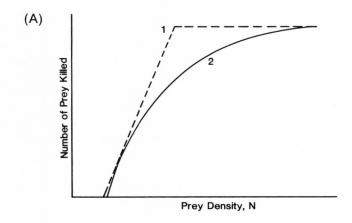

(B)

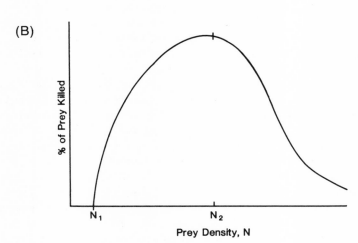

FIGURE 5-4

Total response of predators to prey density. (A) Number killed versus prey density, (1) from Type I functional response, (2) from Type III functional response. Modified from Holling (1959). © Entomological Society of Canada. Used with permission. (B) Mortality rate versus prey density, from Type III functional response. Prey mortality due to predation is density-dependent between N_1 and N_2.

the host's population is killed, resulting in local near-extermination, reducing the host population below N_1 and effectively (locally) eliminating the pathogen. Such cycles of delayed outbreak followed by overcompensating density dependence may be responsible for the cyclic behavior of many insects populations, particularly in forest ecosystems (Anderson & May, 1980; Price, 1983). Chesson and Murdoch (1986) suggested that this may occur regularly among host–parasitoid systems as well, and that equilibrial models (like that of Equations 5-1 and 5-2) apply to a minority of actual interactions observable in the field.

Competition

Competition may occur either among members of a single species or between species utilizing the same resource. Intraspecific competition is a density-dependent process with very little time delay; Milne (1962) described it as "perfectly" density-dependent. In general, as a population increases, resources become limiting, resulting in a lowered birthrate and increasing emigration and death rate. The process is described most simply by the unadorned logistic equation (Equation 4-6, Figure 4-2). Intraspecific competition may simply result in fewer and smaller individuals, or in more overt interactions, including cannibalism, which occurs even in some normally phytophagous insects when they become crowded (Polis, 1981). Cannibalism occurs among many grain-eating insects long before food is in short supply. Some larvae, such as those of the corn earworm or codling moth, cannibalize whenever more than one is present in a single fruit. From a management standpoint, population densities at which intraspecific competition occurs usually exceed the EIL anyway. Except for raising the EIL, there is probably little a pest manager can do realistically to manipulate intraspecific competition in his favor.

Interspecific (between species) competition is assumed by most insect ecologists, yet evidence for its routine occurrence in nature is equivocal (Wiens, 1977; Arthur, 1982; Schoener, 1982). Studies by Paine *et al.* (1981) on bark beetles, by McClure (1980) on hemlock scales, and by Inouye (1978) on bumblebees, among other studies, indicate interspecific competition. However, Rathcke (1976), for example, has shown that extensive overlap in use of resources is not in itself evidence of competition. Coexistence of apparent "competitors" is attributed variously to host density and changing land use (Chew, 1981), to maintenance at lower-than-competitive densities by natural enemies (Faeth & Simberloff, 1981), or to periodic habitat disruption by inclement factors in the physical environment (Andrewartha & Birch, 1954; Lawton & Strong, 1981). Careful studies of insects in the field have often shown that apparent "competitors" differ from one another in subtle ways. For instance, after careful study of a "classic" case of alleged competition between the parasitoids *Aphytis lignanensis* and *A. melinus* (DeBach & Sundby, 1963), Luck *et al.* (1982) found that the two species differed enough in the size of host (California red scale) so as to be not totally competitive. Furthermore, *A. melinus* was more tolerant of extreme temperatures and dispersed more readily (Luck & Podoler, 1985). Competition is usually inferred, but rarely demonstrated, from such differences. Arthur (1982) and Strong *et al.* (1984) concluded that there was little evidence to demonstrate conclusively the widespread occurrence of interspecific competition, especially among phytophagous insects.

Whether it occurs regularly or not, competition as a pest management tool is probably limited to specialized situations, in which competition is for a "neutral" resource (Moon, 1980) rather than for the item that the pest is damaging. It certainly would not do to introduce the large white butterfly *Pieris brassicae* into North America in the hope that it would outcompete the imported cabbageworm *Pieris rapae* in the destruction of cole crops. However, insects such as biting flies whose larvae feed in dung might be managed by introduction of more efficient nonbiting dung-feeders that would outcompete the original fly larvae for this resource (Hughes *et al.*, 1978). The apple rust mite reportedly outcompetes the European red mite for apple foliage (Croft & Hoying, 1977) and has the additional merit of serving as an early-season food source for predaceous mites that later switch to European red mites once the latter become common. Coulson *et al.* (1976) reported that competition from the wood-boring beetle *Monochamus titillator* reduced numbers of the more destructive Southern pine beetle.

Mutualism

Mutualism, like interspecific competition, seems to occur widely in nature, yet is notoriously difficult to demonstrate experimentally (Boucher *et al.*, 1982), apart from a few very obvious examples. Moreover, mutualism has not attracted the same amount of attention on the part of researchers as has been spent on predation and competition. With additional study, it may be possible to reduce pest densities by removing their mutualists. For instance, populations of some aphids might be held in check by natural enemies if their mutualistic ants were removed. Ants attending the treehopper *Vanduzea arquata* on black locust also reduced predation of leafminer eggs by nabid bugs; this activity could lead to higher leafminer populations and damage to the host plant (Fritz, 1983). Similar mutualistic insect–plant interactions might be manipulated for more effective pest management (Chapter 6).

SYNOPTIC MODEL OF INSECT POPULATION DYNAMICS

Southwood (1975, 1977; Southwood & Comins, 1976) proposed a graphical model of insect population dynamics that is helpful in illustrating the interaction of predation and competition with reproduction. It is a general model whose value is primarily instructional, though there are a few field studies whose results fit the model.

The development of the model commences with the relationship between mortality and density. (For simplicity, emigration is included with mortality, and immigration is considered a component of birthrate). Figure 5-5A shows the total predator response (from Figure 5-4B) superimposed on mortality due to competition. The latter increases in perfectly density-dependent fashion according to the classical logistic model. (Note that the horizontal scale is log of population density). Summation of the two curves yields total density-dependent mortality.

Figure 5-5B adds birthrate to mortality. Births obviously are few or nonexistent at exceedingly low N, rise sharply at first, then decline as crowding increases. The shape of the fecundity curve is less important than its relationship to mortality; where fecundity exceeds mortality, the population density increases: $R_0 > 1$ (see Chapter 4). This region is shaded in Figure 5-5B. Where mortality exceeds fecundity ($R_0 < 1$) the population density declines.

This relationship is clarified in Figure 5-5C, which describes the relationship between generations. The solid line represents the relationship between population density in successive generations. Figure 5-5D shows an example: Suppose that the starting population is N_1. If we locate N_1 at point 1 on the graph, we find that the population density in the subsequent generation (N_{t+1}) is N_2. Now, taking N_2 as N_t on the ordinate, we locate point 2, the population density in the next generation, N_3, and so forth, to determine successive densities over eight generations. The results of this exercise are depicted in Figure 5-5E, wherein the successive N_t's are plotted against time.

In the shaded region of Figure 5-5C, $R_0 > 1$ and the population increases; below the line $R_0 = 1$ (where $N_t = N_{t+1}$), the population

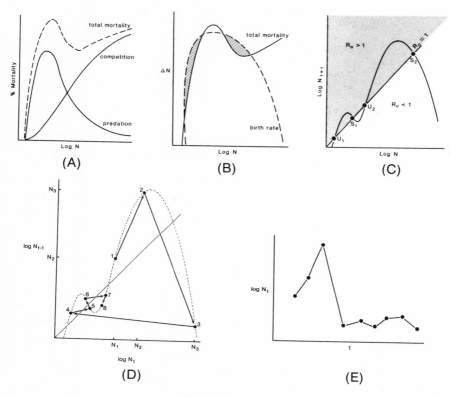

FIGURE 5-5

Synoptic model of insect population regulation. (A) Relationship of mortality rate to log of prey density. Total density-dependent mortality is sum of predation and competition curves. (B) Relationship of population numbers to change in density; $R_0 > 1$ in shaded area. (C) Relationship between log densities of successive generations, with stable equilibria at S_1 and S_2 and unstable equilibria at U_1 and U_2. (D) Example showing density changes over eight generations, starting at point 1. (E) Example of (D) with N plotted over time. (A), (B), and (C) from Southwood (1975). © Academic Press, Inc. Used with permission.

declines. There are four equilibria; S_1 and S_2 are stable equilibria and U_1 and U_2 are unstable. S_1 denotes the point at which predation is effective, whereas S_2 is analogous to K, the carrying capacity of the simple logistic model. U_1 is the same as the minimum population size or "extinction threshold" of the modified logistic (Chapter 4, Equation 4-8 and Figure 4-3), and U_2 is a "release point," beyond which predation is incapable of regulating the population.

The synoptic model has the following consequences:

1. Natural enemies are effective only at densities below U_2. Southwood (and others) called this the "natural enemy ravine,"
   though it has been demonstrated to occur even in the absence of
   predation as a result of differential immigration (Carter &
   Dixon, 1981).
2. The demography of endemic populations ($N < U_2$) may be
   very different from that of epidemic populations ($N > U_2$).
   The latter often are subject to extreme fluctuations, as Figures
   5-5D and 5-5E attest.

The synoptic model is expanded upon below, following further
consideration of the role of pest populations in ecological communities.

THEORETICAL AND ECOLOGICAL ISLANDS

As noted earlier, natural and anthropogenic communities are composed of many species displaying extremely complex interactions involving simultaneous predation, competition, commensalism, and
mutualism. Dealing with this complexity poses not only practical
problems (managing pests in crops without undue perturbation) but
also the theoretical problem of elucidating general principles for analysis and instruction.

Among general theories of community ecology, one that was
received with much initial enthusiasm is the theory of island biogeography, first propounded by MacArthur and Wilson (1963, 1967). In
their view, the number of species sustained by an island ecosystem was
determined by an equilibrium between the rates of immigration and
extinction (Figure 5-6). Support for the theory came both from empirical observations of newly formed islands (such as those formed after
volcanic activity) and from monitoring immigration of the arthropod
fauna of small mangrove islands after fumigation of the initial inhabitants (Simberloff & Wilson, 1969). At the outset, immigration rate is
expected to be high, owing to the paucity of species on the island. As
time passes, the immigration rate declines, simply because chances are
greater that a given species is already present. Conversely, extinction
rate is at first low because so few species are present; as more arrive,
chances increase that any one will become extinct.

The size of the island and its distance from the source of colonists
are expected to alter the slopes of immigration and extinction curves

(A)

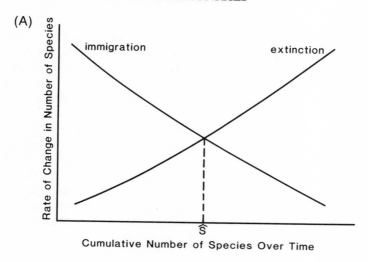

(B)

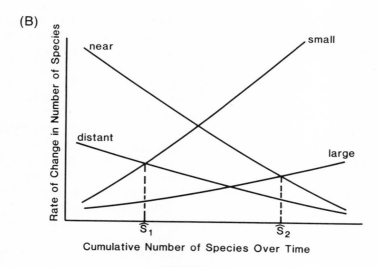

FIGURE 5-6

Relationship between immigration and extinction rates on islands. (A) Relationship of rates to time; equilibrium species number at $\hat{S}$. (B) Variation in rates with respect to size and distance of island. Nearby large island has more species ($\hat{S}_2$) than does distant small island ($\hat{S}_1$). From MacArthur and Wilson (1967). © Princeton University Press. Used with permission.

(Figure 5-6B) so that, in general, small and distant islands have fewer species ($\hat{S}_1$) than do large, nearby islands ($\hat{S}_2$). This again is supported by empirical observation, though there are exceptions, along with competing explanations as to why larger islands support more species than do smaller islands. For instance, in reality larger islands often possess a greater variety of habitats, and the variation in physical environment is likely to be greater.

The MacArthur–Wilson model is a useful generalization though limited in applicability. Rates of immigration and extinction are difficult to document adequately. Simberloff (1976) argued that true immigration should be limited to instances wherein a successful colonist has reproduced and left viable offspring. Many suspected "immigrants" observed by census may simply be resting before continuing on. Similarly, measurements of extinction may be biased by the use of a sampling technique that is simply not sensitive enough to detect a low-density population. The model does not account for species that become extinct, then reimmigrate. Perhaps its most serious practical shortcoming is inability to predict which species will be present, which is an important consideration in pest management.

A corollary of the MacArthur–Wilson theory that has received a large amount of attention and discussion by applied insect ecologists is the concept of r- and K-selection. As originally intended, r-selection referred to selection favoring maximum growth in an uncrowded population, with a resultant suite of adaptations typifying an "r-strategist" (Table 5-1; also MacArthur & Wilson, 1967). K-selection, by contrast, favors competition in a crowded population, resulting in adaptations characterizing "K-strategists." Broadening of the concept by subsequent researchers and writers has caused a great deal of confusion (Parry, 1981) by adding density-dependent population regulation as characteristic of K-strategists, while considering r-strategists to be density-independent. Others have generalized about stability of habitats suitable for r- and K-strategists (Table 5-1). These ideas may have heuristic value in discussions of insect ecology, though (1) the original concept was more limited, and (2) adaptations involve complex trade-offs that mask (or mislead) efforts to categorize insects (or anything else) neatly into "r- and K-strategy". The same insect species often has characteristics of both; for instance, Tallamy and Denno (1982) found that the fecundity of the tingid bug *Gargapha solani* increased in the absence of a guarding female ("r-strategy"), while reduced fecundity led to higher survival ("K-strategy"). Long-winged planthoppers with higher dispersal ability (K) produce fewer eggs than do highly fecund, short-winged morphs (r) (Denno, 1977; also, Chapter 4). Many aphid species produce parthenogenetic, wingless reproductives in response

TABLE 5-1
Comparison of r-Strategy and K-Strategy

Characteristic	Strategy	
	r	K
Fecundity	High	Low
Survivorship	Low	High
Competitive ability	Low	High
Relationship between generation time, T, and habitat stability, S	$T \geq S$	$T \ll S$

to an abundance of nutrients, and produce sexual, winged morphs with much lower fecundity when local food supplies decline in late summer.

Despite these shortcomings in the general concept of r- and K-selection, it is evident that many agricultural pests possess a complex of life history traits consistent with the concept of "r-strategy": short lifespan, great dispersal ability, high fecundity, and (sometimes) parthenogenesis (Stinner et al., 1983). Agricultural crops can be considered "ecological islands," though the analogy must be qualified, for land intervening between suitable habitats for colonization by pest insects is often favorable for survival of dispersing forms. However, agricultural crops, particularly annuals with a short growing season and periodic disruption, represent precisely the sort of environment in which MacArthur and Wilson predicted r-selection to occur. Successful colonists of crop plants thus may display superior powers of dispersal and reproduction, enabling them to locate and exploit small, ephemeral habitats quickly and efficiently. Locally at least, we often observe $R_0 \gg 1$ and r approaching r_m among these arthropods (Chapter 4).

Furthermore, agricultural crops superficially resemble islands. Individual plants as well as monocultural fields are isolated from similar habitat by regions of varying lethality. Price (1976) documented colonization of soybeans by arthropods following planting and found that, in general, herbivores preceded carnivores and that the attractiveness of the crop, its "colonizability," changed with time. I have noted the same tendencies in collards, regardless of the vegetational background into which they have been planted. Annual crops are ephemeral in space and time, and there is insufficient temporal stability for equilibrium to be reached between immigration and extinction (Figure 5-7). Harvest comes too early and the system is reset to zero at the outset of the subsequent growing season. Equilibrium is

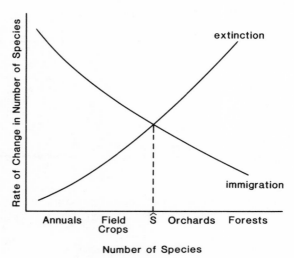

FIGURE 5-7

Speculative relationship of immigration and extinction rates to longevities of crop ecosystems. Short-term cropping systems are unlikely to reach equilibrial $\hat{S}$.

more likely in longer-lasting orchard or forest ecosystems, though it is critical to determine the specific insects present and the EILs for the pests in the system.

Strong (1974; Strong *et al.*, 1977) applied the MacArthur–Wilson theory to the accumulation of insect species utilizing specific crops in different regions of the world. His general conclusion was that the longer a crop had been grown in a particular region, the greater would be the diversity of phytophagous (and carnivorous) insects associated with it (Strong, 1979; see also Southwood, 1961). One implication is that insect pest problems on a crop may decline as an "evolutionary equilibrium" is achieved, and a balance among plant–herbivore–carnivore trophic levels evolves. Pimentel (1961a) termed this "genetic feedback," in which interacting populations of predator and prey select traits in one another that assure their mutual continued existence, fortuitously resulting in the regulation of each population. Obviously, it is critical to know the density (in relation to EIL) at which regulation occurs. Moreover, catastrophic interventions, such as treatment with insecticide, can delay or disrupt coevolution of predator and prey.

SYNOPTIC MODEL WITH r- AND K-STRATEGY

Southwood (1975, 1977; Southwood & Comins, 1976) developed the synoptic model further to account for differences in population regulation between r- and K-selected species. Figure 5-8 replots Figure 5-5C on a three-dimensional graph with the third dimension being the "r–K continuum." Stable population density ($N_t = N_{t+1}$, $R_0 = 1$) is represented by the plane slicing horizontally through the graph. Figure 5-5C represent the "intermediate" strategy of Figure 5-8, and S_1 S_2 U_1, and U_2 are the equilibrium points, as before. Southwood characterized the two humps in the three-dimensional surface as ridges: an "endemic ridge" (U_1 to S_1) and an "epidemic ridge" (U_2 to S_2) separated by the natural enemy ravine (S_1 to U_2), the only densities at which predation effectively regulated N (as in Figure 5-5C).

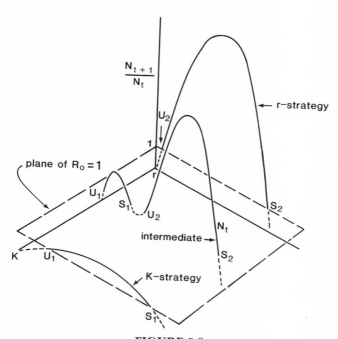

FIGURE 5-8

Synoptic model in three dimensions; "r–K continuum" has been added to Figure 5-5C. Dashed lines lie beneath the plane of $R_0 = 1$. From Southwood and Comins (1976). © British Ecological Society. Used with permission.

In the demography of an extreme r-strategist, one whose powers of reproduction and dispersal consistently overwhelm any significant impact of predation, the population curve is entirely on the "epidemic ridge," with marked changes in density from generation to generation, a consequence of the high value of r. This is illustrated by Figure 5-9A, in which the demography of the extreme r-strategist is shown in two dimensions, comparable to Figure 5-5C. Such a population is prone to wild fluctuations, as is clear from Figures 5-9B and 5-9C. The demographies of some aphid species and of phytophagous mites approach this at times.

On the contrary, the demography of an extreme K-strategist lies exclusively within the endemic ridge, with N_{t+1} versus N_t between U_1 and S_1 (Figures 5-8, 5-10A, 5-10B, 5-10C). Fluctuations are of far lower amplitude, a result of the relatively low r. (Analogous results can be obtained from the simple logistic model, Figure 4-6; recall that the

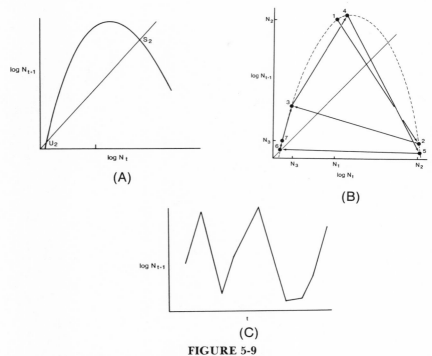

(A)

(B)

(C)

FIGURE 5-9

Synoptic model for extreme r-strategist. (A) Relationship between densities of successive generations (compare with Figure 5-5C). (B) Density changes over eight generations, starting at point 1 (compare with Figure 5-5D). (C) Same as (B), with density plotted over time (compare with Figure 5-5E).

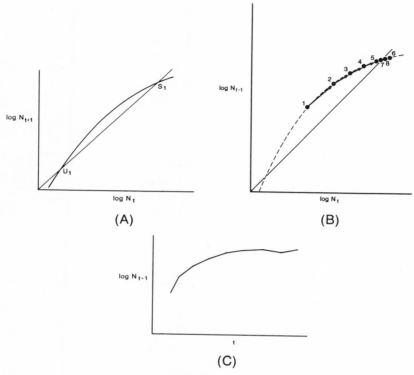

FIGURE 5-10

Synoptic model for extreme K-strategist. (A) Relationship between densities of successive generations (compare with Figure 5-5C). (B) Density changes over eight generations starting at point 1 (compare with Figure 5-5D). (C) Same as (B), with density plotted over time (compare with Figure 5-5E).

greater the r, the wider and more chaotic were the population fluctuations.)

An important consequence of the synoptic model is that natural enemies might have little impact on either extreme of the "$r–K$ continuum." r-strategists simply outbreed predation, whereas K-strategists presumably would defend successfully against predation, and most would survive at least to reproduction. Southwood (1977) suggested that efforts in "classical" biological control should be directed against pests of lifestyle intermediate between r and K. He proposed that the relative effort expended on various management techniques should approximate that of Figure 5-11. Stenseth (1981), starting from the equilibrium theory of island biogeography, came to an alternative

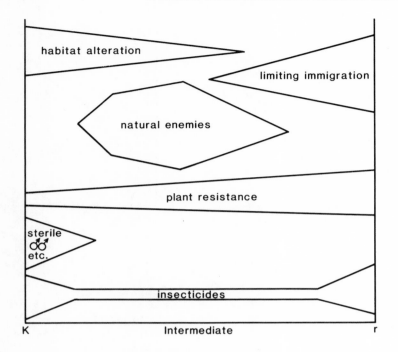

Adaptive Strategy "Continuum"

FIGURE 5-11

Proportion of regulation likely to be provided by selected management strategies in relation to adaptive strategies of insect pests. The width of each plot
represents the approximate contribution to regulation of the pest in question.
From Southwood (1977). © Society of the Sigma Xi. Used with permission.

conclusion. He suggested that, when possible, control may best result
from simultaneously minimizing the rate of immigration and expected time to extinction, and that greatest success may be expected
with methods (such as chemical insecticides) that reduce overall survivorship in K-selected species.

The synoptic model is useful in instruction, although demography of pest populations in the field can be considerably more complex. Numerous studies of insect populations have shown a decided
tendency toward fluctuations with periodic outbreaks reminiscent of
Figure 5-5E (Price, 1983). Outbreaks of aphids (and other insects) can
be induced by application of a selective chemical poison (Horn, 1983),
effectively removing the natural enemy ravine, resulting in epidemic

demography. Karban (1982) showed that periodical cicadas over-whelmed and thereby "escaped" predators at high densities. South-wood and Comins (1976) noted that the population curve for cereal aphids approached the bimodality of Figure 5-5C and Figure 5-8, though Carter and Dixon (1981) found the "ravine" in the absence of significant predation and attributed it to the vagaries of intermittent migration. Natural enemy ravines (though not termed such) have been demonstrated in many forest Lepidoptera subject to periodic outbreaks (e.g., spruce budworm and fall webworm; Morris, 1972). The precise causes for periodic outbreaks and population cycling are unclear in many cases and probably vary from one system to another, even in the same species. In a general model reminiscent of Southwood's, Berry-man (1982) defined an "epidemic threshold" (analogous to U_2 in the figures) that a population may exceed through a chance, stochastic, event. Perhaps a windstorm may increase suitable habitat for wood-boring beetles, or a change in market conditions may prompt a vast increase in acreage of a formerly unprofitable crop (such as soybeans in the midwestern United States recently). For pest management to be effective, it is imperative to quantify and predict, as far as possible, the events preceding epidemic outbreaks of insects and other plant pests.

DIVERSITY AND STABILITY

Much has been made of a purported relationship between species diversity and population stability (Goodman, 1975; Kimmerer, 1984). Supposedly, complex ecosystems, having more trophic interconnec-tions in their food webs, are thought to be buffered against major density changes within any component species. Figure 5-12 shows a few imaginary communities to illustrate this point. The three-species community (Figure 5-12A) is more susceptible to changes in density of the herbivore H1 than is the nine-species community (Figure 5-12B). In the extreme, suppose herbivore H1 was exterminated locally; in the three-species community, carnivore C1 would also be eliminated and plant P1 might increase dramatically, whereas these events would be less likely if herbivore H1 were lost from the nine-species community. Many additional trophic links are possible in the nine-species commu-nity, as Figure 5-12C attests; if the "carnivores" are sufficiently omniv-orous, there may be as many as 33 trophic links, excluding cannibal-ism.

Species diversity is usually measured by an index that combines numbers and relative abundances of the species in a community, as a reflection of the number of trophic links. Numerous indices of diver-

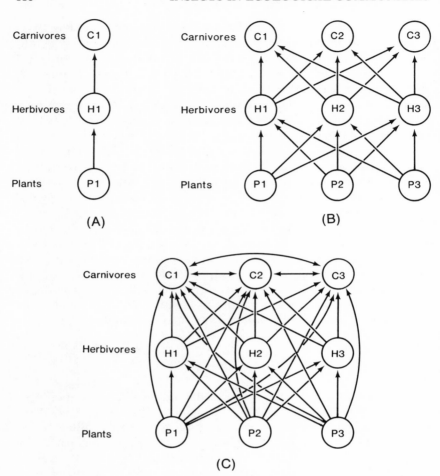

FIGURE 5-12

Structure of hypothetical communities. (A) Three-species community in which loss of any one species may cause serious perturbations throughout remaining two populations. (B) Nine-species community in which loss of any species will have less severe overall effect than in (A). (C) Same as (B), with broader diets for "carnivores"; addition of links is predicted to buffer the system even further against perturbations.

sity have been proposed by ecologists, and Southwood (1978), Price (1984), and Pielou (1977) have reviewed most of these. The most popular is the Shannon–Weaver index:

$$H = -\sum p_i \log (p_i) \qquad (5\text{-}3)$$

where p_i is the proportion of the total sample made up by each species. Kempton and Taylor (1976) criticized this index and some others on the basis that unjustified weight is given to very common and very rare species. They proposed, rather:

$$Q = S/2 \times \log (R_2/R_1) \qquad (5\text{-}4)$$

where S is total number of species in the sample, R_1 is lower quartile of species abundance distribution, and R_2 is upper quartile of species abundance distribution.

Of course, it would be best to measure the actual number of trophic links; the nine-species community of Figure 5-12 would yield an identical diversity index, despite having anywhere from 6 to 33 links.

Diversity indices are useful shorthand for comparing entire communities, yet it is an open question as to whether a general relationship actually exists between species diversity and ecosystem stability, and, if it does, what the cause of this relationship is and whether it is of any use in insect pest management. Results of field experiments have been equivocal (Chapter 6), and it seems that as much depends on the specific relationships among the species involved as on any general emergent properties of the community as a whole.

Agricultural ecosystems in particular display rather loose trophic interconnections, and urban ecosystems are even more loosely "organized." Researchers are therefore justified in using caution in applying principles derived from study of natural ecosystems of many centuries' standing to the artificial assemblages that comprise most human-dominated ecosystems. However, the insight to be gained from study of natural systems and from simple, general population models is most useful in devising complex and specific models essential to modern pest management (Chapter 12). The interrelationship of plant diversity and insect populations is discussed further in the next chapter.

CHAPTER

6

INSECT–PLANT INTERACTIONS

Strong *et al.* (1984) estimated that one fourth of all species of animals and plants (and therefore one third of all species of insects) are phytophagous insects. It is remarkable that so few species in this bewildering array have become agricultural or silvicultural pests, yet the depredations of these several hundred pests have been troublesome and occasionally disastrous. Probably, humans have been aware of intimate relationships between insects and plants since before the dawning of agriculture. In a sense, the overall strategy of agricultural pest management has been and remains the reduction of the impact of phytophagous insects so that the largest possible proportion of a commodity can be harvested for human benefit.

Only recently have the subleties of insect–plant interactions become apparent, especially an appreciation of chemically mediated interactions between insects and their host plants. Fraenkel (1959) early articulated the putative role of these "secondary" chemical substances in plant defense against insect attack. Ehrlich and Raven (1964) extended Fraenkel's observations to develop the notion that phytophagous insects and their host plants represented coevolved systems, though the extent of coevolution and the intensity of selection remain subject to debate (see Jermy, 1984). Research during the past two decades has revealed that these presumably defensive chemicals are far more prevalent in plants than was earlier assumed. A thorough appreciation of the insect–plant chemical interface offers exciting prospects for insect pest management, especially through the development of resistant varieties (Chapter 10). Discoveries and fresh hypotheses appear regularly in this rapidly developing field, and one might expect that some of the ideas that you read herein may soon be modified in the light of new information. Moreover, the diverse array of adaptations represented among the hundreds of thousands of phytophagous insect species, and the variable experimental conditions under which they are studied, make it difficult to formulate valid generalizations.

This chapter commences with a consideration of insect impact on plants, followed by discussion of plant responses to insect attack. Spacing, environmental "texture" (as defined by Root, 1975), and mono- versus polyculture are also discussed, as they are important to an understanding of insect–plant interactions. For further consultation, Denno and McClure (1983), Strong *et al.* (1984), and Miller and Miller (1986) are useful recent syntheses. (Each contains a thorough bibliography.)

INSECT IMPACT ON PLANTS

Phytophagous insects may be broadly considered to exhibit one (or more) of the following four feeding types.

1. Chewing insects, such as Orthoptera, Coleoptera, and many larval Lepidoptera, chew holes in leaves, buds, stems, and other plant parts.
2. Sucking insects (mostly Homoptera and Hemiptera) remove sap and are major vectors of plant pathogens.
3. Mining insects (primarily, larval Diptera, Lepidoptera, and Coleoptera) bore in leaves and stems.
4. Galling insects feed on liquids within the protection of swellings they induce in plant tissue.

Any given plant species, including agricultural and ornamental crops, may harbor insects of all four feeding types, representing a melange of specialists feeding only on that plant species, and opportunists with broad feeding preferences. Members of any feeding group may become pests.

Insects' impacts on their host plants may be both physical and chemical, and the interplay between the two is often inseparable. The most obvious impact is the clearly visible injury; holes in leaves, wilting, and curling and/or galling are well-known to anyone with more than a rudimentary knowledge of the natural world. However, "damage" as defined by farmers or homeowners may not represent major biological dysfunction, and applied entomologists need to consider this when formulating EILs. Many plant species are well-adapted to withstand light-to-moderate defoliation with no loss of yield. In an average season, forest trees are usually 10% to 15% defoliated anyway, and Kulman (1971) pointed out that deciduous forest trees can withstand up to 30% defoliation over three consecutive years without mortality or even measurable reduction in growth of wood. (However, conifers are much more susceptible.) Soybeans tolerate up to 40% prebloom defoliation, and 25% postbloom defoliation with insignificant reduction in yield. Normally, to tolerate such levels of defoliation, plants must be healthy at the outset, with adequate moisture, fertilization, and relative freedom from pathogens. Stark (1965) reviewed numerous studies showing that insect density and injury were reduced in well-fertilized forests. Stressed plants are often more susceptible (see below), and some insect population outbreaks initially suspected of having been caused by favorable weather apparently were induced at least partially by plant stress (White, 1976, 1984). The spruce budworm is an example; periodically, epidemic budworm populations defoliate (and destroy) thousands of hectares of spruce and fir in northern forests of the United States and Canada. These outbreaks follow closely upon drought, fire, or windstorm, and seem to occur when and where the trees have been recently stressed by these density-independent events (Morris, 1963; Rhoades, 1983).

Measurements of the tolerance of plants for defoliation may underestimate or overestimate the actual impact of chewing insects if these studies depend upon mechanical defoliation (Hammond & Pedigo, 1982; Higgins et al., 1984). Feeding by phytophagous insects of several species in low numbers has been shown to stimulate plant growth, due to analogues of plant growth hormones within the insects' saliva. Growth-promoting substances such as indolacetic acid (controlling elongation of plant shoots) are secreted by mandibulary glands of several insect species (Hori, 1974, 1976). Insect feeding may also induce greater production of the plant's own growth-regulating chemicals, or may induce the production of defensive compounds (Haukioja & Neuvonen, 1985; Karban, 1986).

Insect attack may change plant suitability for one or several growing seasons thereafter. Reduced vigor may follow high herbivore densities; Klopfenstein (1977) showed that even moderate densities of the European red mite on apple foliage in June reduced the fruit set the following year. Apparently, the trees were responding to stress by increasing vegetative bud formation at the expense of floral buds. Alternatively, McClure (1983) showed that high densities of hemlock scales reduced foliage quality sufficiently to cause a decline in scale density during years following outbreaks. Heavy attack by phytophagous insects may reduce the density of their specific hosts, thereby reducing competition from dominant plant species and allowing greater local diversity (Caughley & Lawton, 1981).

As a general rule, moderate-to-severe insect attack is sufficiently stressful that plants become more susceptible to other environmental hazards. Wood-boring beetles weaken their hosts structurally, thereby rendering them more likely to be damaged or destroyed by drought, fire, or storms. The beetles also provide sites of entry for pathogenic microorganisms. Phytophagous insects may spread pathogens among either stressed or vigorous plants. (The role of insects in vectoring plant pathogens is a major area of study that would require a separate chapter at least. See Carter, 1973, Maramorosch and Harris, 1981, and Harris and Maramorosch, 1982, for details and entry into the literature of insects' vectoring plant pathogens.)

A few insects feed little or not at all, yet they damage plants during oviposition. Cicadas chisel oviposition slits into twigs of hardwood trees and large numbers of periodical cicadas may devastate apple or peach orchards in the eastern United States. Once the eggs hatch, nymphs drop to the ground and burrow to the roots. It is presumed that the slashing of twigs stresses the tree and enhances nutrient transport within the roots, to the benefit of the cicada nymphs' nutrition. This has yet to be substantiated.

PLANT EFFECTS ON INSECTS: NUTRITION

It is elementary that plants provide water and nutrients to phytophagous insects. The amounts of these materials that insects obtain from plants therefore may limit the insects' survival or growth. Even water may be a limiting factor for insect herbivores.

Generally, it is nitrogen (in the forms of amino acids and proteins) that is the plant nutrient most likely to be in short supply for insects (Price, 1984). Nitrogen content of plant tissues and sap is much lower (about 1–2%) than in the insects themselves, and many studies have shown a clear relationship between nitrogen availability in plants and fecundity of their insect fauna. Seasonal pulses in aphid reproduction are one example (van Emden et al., 1969). Availability of nitrogen also affects the growth rate of immature insects, and its deficiency can lead to high juvenile mortality (White, 1978).

Phytophagous insects respond with a number of adaptations that enhance their efficiency in nitrogen removal from plants. Feeding is often concentrated at the growing point of the plant and on ripening reproductive parts, which generally contain the highest concentrations of nitrogenous compounds. These same floral or fruiting structures are high in secondary "defensive" chemicals also. The availability of nitrogen in growing tissues is apparently a strong selective pressure resulting in early-season feeding despite the rigor of unpredictable weather conditions (Feeny, 1970). Many apparent herbivores are not 100% phytophagous but will feed occasionally on other insects or insect eggs, and may rarely take vertebrate blood. (The reader who is unconvinced may verify this with third- or fourth-instar *Lygus* bugs.)

Fertilization of crop (and other) plants therefore might be expected to increase the incidence of insect attack, though results of experiments to date have been equivocal. Hargrove *et al.* (1984) found that fertilizing black locust trees resulted in higher initial herbivore densities but that densities declined with time as the plants apparently developed resistance. They suggested that the time at which populations were censused affected the outcome of fertilization studies. Denno (1986) found that heavily fertilized marsh grass led to production of larger planthoppers with higher fecundity. In my own research I consistently find that aphid populations with the highest reproductive rates are on the largest and leafiest collard plants.

Other agricultural practices may influence plant nutrition and through it, herbivore populations, and these impacts must be considered carefully in the development of pest management programs. Application of the herbicide 2,4-D increased nitrogen in corn, resulting in a threefold increase of corn leaf aphids and a 26% increase in corn borer populations (Oka & Pimentel, 1974). Application of chlorinated hydrocarbon insecticides to corn can stimulate increased synthesis of amino acids (Thakre & Sazena, 1972). Irrigation of corn increases populations of western corn rootworm (Hill & Mayo, 1980), and irrigated cotton harbors many pests in its lush foliage (Chapter 12).

PLANT EFFECTS ON INSECTS: DEFENSE

Plant chemical defenses are generally categorized as quantitative or qualitative (Feeny, 1975), though there is some debate over the utility of these terms. Quantitative defenses are effective against nearly all herbivores and are considered typical of plants in later stages of ecological succession, when a premium is placed on long-term survival. Conversely, qualitative defenses prevail among early-successional plant species whose rapid growth and reproduction are favored and less energy is diverted into the manufacture of secondary chemical substances. Such qualitative defenses are effective against a more limited variety of herbivores, and it is considered more likely that herbivores may evolve specializations to avoid or overcome qualitative

defenses. Futuyma (1976) and Price (1984) noted that a higher proportion of herbivores feeding on early-successional plant species were
specialists, as compared to the more "generalist" herbivores on plants
typical of later successional stages. Most agricultural crops are annuals
derived from early-successional ancestors (Chapter 5), and typically
each major family of agricultural crop plants has its own assemblage
of phytophagous insect specialists. The categories "generalist" and
"specialist" are arbitrary, definable with respect only to one another,
as is the case with r- and K-strategists (Chapter 5). For the present
discussion, I consider generalists to be species feeding indiscriminately
upon members of more than two unrelated plant families (and usually
on many families; e.g., black cutworm, *Agrotis ypsilon*), whereas
specialists normally feed upon members of a single plant family (e.g.,
Colorado potato beetle on Solanaceae). Many generalists are specialists
within each localized region (Fox & Morrow, 1981). Scriber (1983)
scrutinized several "generalist" Lepidoptera and showed that locally
they were often specialists. They fed upon a variety of forest trees
throughout their entire geographical ranges, but specialized on a very
few tree species in each specific locality. These physiological and
detoxification abilities have a genetic basis indicating local adaptation
to specific host plants.

Plant suitability for insect hosts is much more variable than had
once been assumed. Individual plants of the same species may differ
greatly one from one another in their defense against insect attack.
Alstad and Edmunds (1983) found that ponderosa pines within a
single local stand differed markedly from one another in the density of
black pineleaf scales, and that scale populations on each tree were
genetically distinct from those on neighboring trees. In this way the
scales apparently overcame the host's defenses, at least partially, but at
the cost of reducing their effectiveness at colonizing and exploiting
other host individuals. Denno and McClure (1983) suggested that such
interplant variation is a defense in itself, preventing herbivores from
evolving "fine-tuned" defenses against long-lived plants. Leaves on
the same tree—indeed, on the same twig—may display up to four-
fold differences in concentrations of various secondary compounds
(Schultz, 1983). Hungry caterpillars pass over the less attractive leaves
(of maple and birch, at least), and concentrate their feeding on the
more delectable ones. Leaves also change chemically with age, and
temperature may affect the production of secondary compounds.
Leaves receiving full sunlight are generally less attractive than those in
shade (Raupp & Denno, 1983), though their nitrogen content may be
higher (Mooney & Gulmon, 1982), and they are of course warmer,
allowing greater feeding activity despite reduced chemical attractive-

ness. Information on these aspects of host plant "quality" (Scriber, 1984) must be added to models of population dynamics for phytopha- gous insects if these models are to be expected to mimic accurately the natural world. A simple way to add this into the models introduced earlier is to consider the growth rate, r, of Equation 4-7 to be a variable function dependent on host plant condition. Luck and Scriven (1979) used this approach in modeling elm leaf beetle populations. Jansson and Smilowitz (1986) found increased growth rate (r) among green peach aphid populations in response to enriched nitrogen in potatoes. In general, the more favorable the host, the higher is r. In terms of the synoptic model (Chapter 5), a highly favorable host may raise the entire curve N_t versus N_{t+1} to eliminate the "natural enemy ravine."

Host suitability may change seasonally and in response to stress- ors such as other insects, pathogens, and drought. Generally, it is early in the growing season that plants are most susceptible to insects. Young, succulent leaves contain higher concentrations of qualitatively defensive chemicals and lower concentrations of quantitative com- pounds (Scriber, 1984), and most energy is directed toward growth. Younger leaves are richer in the nutrients necessary to sustain the attacking insects. Feeny (1970) showed that oak leaf tannins accumu- lated throughout the growing season, so that older foliage had far less available nitrogen. As a consequence, most common species of Lepi- doptera fed on young leaves, early in the season. However, senescent foliage also can be suitable and may be preferred by phytophagous insects; Trumble (1982) showed that green peach aphids' growth and survivorship were higher on the oldest leaves of cabbages, for there was a lower concentration of defensive chemicals, and nutrients (especially nitrogen) were more readily available, due to the chemical breakdown occurring with the leaves. Ikeda et al. (1977) found that young foliage of pines deterred feeding by sawfly larvae, concentrated on older fo- liage.

Physical Defenses

Plants possess a variety of adaptations to overcome insect attack, and some of these are physical characteristics that reduce a plant's attrac- tiveness or nutritive value. Raupp (1985) found that mandibles of the leaf beetle *Plagiodera versicolora* became worn after the beetles had fed on older and therefore tougher leaves of willow, resulting in reduced feeding and, consequently, lower fecundity. In other cases, there may be actual insect mortality caused by the plant (Norris & Kogan, 1980). For instance, the tiny spinules on leaves of soybeans (and other plants) can be lethal to leafhoppers (Pillemer & Tingey, 1976). Alternatively, plants may simply abscise the affected portion, as Whitham (1983) noted for cottonwood leaves containing galls of *Pemphigus* aphids. A large proportion of plum curculio larvae meet a premature death due either to early drop of fruit (Levine, 1977) or to being squashed within the enlarging fruit, especially when in a host (such as apple) to which the curculio is not well-adapted. Resins secreted by conifers and *Prunus* spp. are effective at healing wounds quickly and entrapping wood-boring insects and bark beetles. Plants may simply overwhelm local populations of seed predators by producing abundant crops at infrequent intervals (e.g., Janzen, 1971). Beattie *et al.* (1973) showed greater synchrony of *Lupinus* flowering in the presence of phytophages than when the insects were absent. In such an overabundance, there is an ample food supply for caterpillars, yet many flowers still escape to bloom and successfully set seed.

Chemical Defenses

As noted earlier, many species of green plants possess secondary metabolic products, chemicals that are not involved in primary metabolism (Whittaker & Feeny, 1971; Rosenthal & Janzen, 1979). For a long time, the adaptive significance of these "allelochemical" compounds was a subject of controversy, because they seemed to play no role in routine nutrition of the plant. More recently, it has become clear that in many if not most cases, one major function of these plant products is the defense of the plant against insect attack, though they have other biochemical functions (Chew & Rodman, 1979). Many allelochemicals give distinctive flavors to members of certain plant families, and are the source of many culinary spices and herbs that enhance our enjoyment of eating. (Doses for human mortality are a great deal higher than those for insects, despite young children's occasional complaints that all vegetables are poisonous.) A few of these chemical compounds have been extracted for use as insecticides for over a century (e.g., nicotine from tobacco, pyrethrum from some chrysanthemums).

Allelochemicals reduce insect survivorship and/or fecundity by poisoning, and some are also antifeedants or repellents. There is considerable variety in their mode of action. Nicotine and pyrethrins are neurotoxins (Chapter 7). Xanthotoxins from Umbelliferae interfere with proper replication of DNA under ultraviolet light (Berenbaum, 1978). Leaf tannins combine with proteins and may render them indigestible to caterpillars (Feeny, 1970), though this is not certain (Berenbaum, 1983). Growth of some acridid grasshoppers is enhanced by similar tannins (Bernays & Woodhead, 1982), and allelo-chemicals may be nutrients for some insects (Scriber, 1984).

Sláma (1979; Sláma & Williams, 1965) isolated from balsam fir a compound with juvenile hormone activity against pyrrhocorid bugs, and since then insect hormone analogues have been isolated from a number of plant species from several orders (e.g., Bowers & Nishida, 1980; Nishida et al., 1983). Other plants possess compounds ("preco-cenes") that interfere with the insects' own juvenile hormone (Bowers, 1981). Bracken fern is particularly high in concentration of molting hormone analogues (Jones, 1983), despite which it supports a fairly extensive fauna of phytophagous insects. Apparently, its "defensive" chemicals are ineffective, which is an indication that some insects are tolerant of the chemicals or that their own physiological systems are capable of overcoming any response to the chemical (Reese, 1979). Locusts (Acrididae) are generally unaffected by hormone analogues, and Rhoades (1983) reported that gibberellin, a plant growth regula-tor, stimulated the growth of locusts as well.

The quantity of allelochemicals varies even within a single plant, and this may be one reason why some leaves are extensively chewed upon while neighboring leaves remain unaltered. Generally, concen-trations are higher in reproductive parts such as flowers and seeds (as is well known to connoisseurs of marijuana and other psychoactive shrubbery). Examples are the high concentrations of pyrethrins in the heads of chrysanthemum, and hypericin in the seeds of Hypericum (Rees, 1969). These reproductive structures are richest in nutrients, especially nitrogenous compounds. The concentration also varies sea-sonally, usually being greatest at the middle of the growing season, at least in temperate climates (Krischik & Denno, 1983). This is often reflected in a greater density and diversity of herbivores early or very late in a growing season. That allelochemicals are indeed effective against insect attack is evidenced by observations such as that there is a greater quantity and diversity of allelochemicals in lupines where Glaucopsyche lygdamus is present than where this butterfly does not occur (Dolinger et al., 1973). In later stages of ecological succession, plant species lacking allelochemicals are generally more palatable and are attacked by a wider variety of insects (Reader & Southwood, 1981).

Feeding by phytophagous arthropods can enhance the chemical defense of plants by inducing production of allelochemicals, though the extent of the phenomenon is hotly debated (e.g., Edwards & Wrat-ten, 1985; Neuvonen & Haukioja, 1985). Cell wall fragments induce production of proteinase inhibitors in response to feeding by the Colorado potato beetle (C. A. Ryan, 1983). Karban and Carey (1984) found lower densities of spider mites on cotton plants that were previously exposed to mites than on unexposed plants, and this resistance persisted after treatment (Karban, 1986), though the biochemical basis was not determined. Tallamy (1985) found that increased fitness in squash beetles resulted from the beetles' behavioral trait of trenching leaves prior to feeding, thereby reducing translocation of defensive cucurbitacins. The intriguing possibility exists that insect feeding might stimulate a defensive reaction with neighboring plants (Baldwin & Schultz, 1983), the so-called "talking trees" hypothesis. Conclusive evidence of this is lacking thus far (Fowler & Lawton, 1985).

The overall impact of insects' feeding on plant fitness is a matter of some debate. In some cases plant reproduction increases in response to herbivory (e.g., Stinson, 1983), while in other instances reproduction decreases under pressure from herbivores (e.g., Marquis, 1984). Whitham and Mopper (1985) demonstrated that feeding by *Diocycta albovitella* resulted in 30% reduction in growth and a fivefold increase in cone production on pinyon pine; this is certainly enough to impact fitness. It is an open question as to whether plant chemical defenses and their insect antagonists have actually coevolved according to the generally accepted scheme of Ehrlich and Raven (1964), or whether specialization by phytophagous insects has evolved in response to other factors (Jermy, 1984). Biological control of weeds is more effective against asexual species, and this suggests that lack of genetic recombination inhibits the evolution of a response to insect attack (Burdon & Marshall, 1981). One often finds closely related insect species feeding upon distantly related plants, and vice versa (Futuyma & Gould, 1979). Jermy (1984) believed that, were the coevolutionary scenario valid, the degree of relationship among insects on the same plant host species would be much closer. He suggested that insects may specialize due to environmental predictability and the need for efficient recognition of host plants, rather than because of evolutionary pressure to overcome plant defenses. Isolated but easily recognizable (chemically) host plants also may serve as easily located mating sites (much as ticks locate their warm-blooded hosts before mating, then mate on them). Strong *et al.* (1984) suggested that equilibrium population densities of most phytophagous insects are normally too low to have significant impact upon the fitness of their host plants.

Effects on Predators and Parasitoids

The interactions between phytophagous insects and their host plants can have an impact on activity and effectiveness of natural enemies. Natural enemies—especially, parasitic Hymenoptera and insectivorous birds—often concentrate their searching activities on areas of visible insect damage, such as holes and ragged edges of leaves chewed upon by caterpillars. Time spent feeding is time spent exposed, and increases these caterpillars' risk of predation (Heinrich, 1979; Weseloh, 1981). The greater the variability in chemical content of leaves (see above), the longer a caterpillar may search for favorable food and the longer it is exposed to predators. Moreover, some allelochemicals slow the growth of herbivores, again increasing exposure time to predation. Predators and parasitoids may use chemical cues from plants to locate potential meals. *Diaeretiella rapae*, a parasitoid specializing in aphids on Cruciferae, orients to volatile mustard oils (Read *et al.*, 1970). Vinson (1975) showed that the parasitoid *Cardiochiles nigriceps* was attracted to "odors" released by the feeding of its host on corn. Sauls *et al.* (1979) found greater kairomone production by *Heliothis zea* larvae when they fed upon living plants than when they were reared on an artificial diet. Phytophagous insects thus may avoid some predation by switching to a different host species (Hendry *et al.*, 1976).

Alternatively, physical or chemical characteristics of the plant may inhibit the effectiveness of parasitoids or predators. Hairy varieties of cucumbers, while somewhat more resistant to Homoptera than are smooth cultivars, also inhibit effective searching by *Encarsia formosa*, the principal parasitoid of the greenhouse whitefly. Campbell and Duffey (1979) showed that tomatine, extracted from tomatoes, inhibited the growth of *Heliothis zea* larvae but was actually lethal to its primary parasitoid, *Hyposoter exiguae*. Smiley *et al.* (1985) found that the leaf beetle *Chrysolina aenicollis* increased its feeding rate on willow leaves high in salicilin content, and apparently predation was reduced. This phenomenon may be more common than the few documentations to date indicate, and there is some concern about the compatibility of biological control with the breeding of resistant varieties of plants (Chapter 10). Other parasitoids may be unaffected by exposure to secondary plant compounds. A tachinid fly parasitizing the monarch butterfly caterpillar sequesters the cardiac glycosides that protect the monarch from bird predation, and the tachinid is likewise protected. Morrow *et al.* (1976) noted a similar instance of sawfly larvae sequestering the noxious oils of *Eucalyptus* spp. and later spitting them upon attacking birds. *Aphytis* spp. do not parasitize California red scales when the latter are on palms rather than on citrus.

Host plant quality may be intimately related to the incidence of parasitism. Price (1984) found that cecidomyiid galls on willow were larger where there was more moisture available, and that these larger galls were too thick for parasitoids' ovipositors to penetrate. Luck and Podoler (1985) noted a direct relationship between scale size and host quality, and found that scales on smaller-sized hosts were simply too small to allow the parasitoid to develop to adulthood.

PLANT SPACING AND INSECT DENSITY

The dispersion of host plants bears an intimate relationship to densities of infesting phytophagous insects that is only beginning to be appreciated. Root (1975) and Kareiva (1983a) referred to this spacing of plants and their relationship to surrounding plants of different species as "texture." The apparent texture of host plants may limit activities of parasitoids and predators or may enhance or inhibit the movement of phytophagous insects themselves by altering their behavior. Host plants surrounded by those of another (and distantly related) species may be much less attractive to colonizing specialist herbivores (Root, 1973; Risch, 1981; Horn, 1981, 1984).

In order to utilize a plant host, an insect must first locate it. Rausher (1983) and others have suggested that an insect species first locates the appropriate habitat, then the host species, and finally a suitable individual of the host on which to feed or oviposit. The cues for each activity may be unique, and "successful" plants may be those that effectively "hide" from insect attack simply by being difficult for insects to locate. It is not easy to discern accurately the appropriate cues used by an insect to locate its host plant. Most such studies have been undertaken in laboratories where conditions are highly artificial. The stimulus used to locate a host in the laboratory may not be that used to locate the host afield, and other factors may complicate normal host-finding behavior. Leaves in the shade may be less likely to be colonized by phytophagous insects (Risch, 1981). Chemical variation may discourage colonization (Jones, 1983). The size of a patch of plants may determine the outcome of colonization; generally a species–area curve (Chapter 5) results. There is thus a danger inherent in extending the results of small-plot studies to large agricultural fields. Kareiva (1983b, 1985) showed that flea beetles (*Phyllotreta* spp.) tended to wander away from very small collard patches, so that their population densities became greater in larger-sized plots. The degree of contagion among plants themselves may influence colonization; Cain *et al.* (1985) showed that collards planted in a regular grid were more easily found by larvae of the imported cabbageworm than were clumped collards at the same density. Plant quality may be altered by crowding, and outbreaks may be confined to "richer" food plants (White, 1976). McClure (1983) demonstrated that fertilization of hemlocks resulted in much higher densities of specialist scales. A very large yet very dense stand of plants may in effect "dilute" the impact of insect response, so that the pest population intensity is lowered.

TEXTURE AND COMMUNITY STABILITY

An ongoing debate among researchers of insect–plant interactions centers around whether increasing plant diversity brings about an analogous reduction in population densities of phytophagous insects (primarily, pest species). Attempts to demonstrate such a relationship have produced equivocal results. For instance, Altieri *et al.* (1977, 1978) found reduced populations of leafhoppers and army worms when corn and beans were interplanted as compared to when each was planted separately. Bach (1980) found that monocultural stands of cucumbers supported 10 to 20 times the density of striped cucumber beetles than did plants mixed with corn and broccoli even when both plantings (of cucumbers) were at the same density. In laboratory experiments the beetles preferred to feed on leaves from cucumbers grown in monoculture (Bach, 1981), raising the interesting prospect that polyculture may enhance production of allelochemicals. Root (1973) noted that collards grown in solid blocs developed higher population densities of all crucifer-eating arthropods (caterpillars, flea beetles, and aphids) than did collards planted in single rows within an old-field habitat. From his results he proposed the "resource concentration hypothesis":

> herbivores are more likely to find and remain on hosts that are growing in dense or nearly pure stands; the most specialized species frequently attain higher densities in simple environments. As a result, biomass tends to become concentrated in a few species, caus-ing a decrease in the diversity of herbivores in pure stands. (p. 95)

Observations of other natural and anthropogenic ecosystems seem to refute a general diversity–stability relationship. Denno (1977) found long-term temporal stability of herbivores in salt marshes composed of only two plant species (*Spartina* spp.), a natural analogue to monocul-ture. Way (1977) suggested that the periodic outbreaks of Rocky Moun-tain locusts that plagued 19th-century settlers on the western plains of the United States were generated partly in response to the diversity of the native grassland ecosystem. Removal of buffalo, large mammalian predators, and rodents apparently created a more simplified system in which (for as-yet-unknown reasons) locusts were no longer prone to outbreaks. Ehler (1978) observed increases in populations of strawberry whiteflies in mixed plantings as compared with pure stands. Forests, traditionally regarded as highly diverse ecosystems, are in many instances prone to periodic outbreaks of borers, bark beetles, and defoliating Lepidoptera, such as the gypsy moth. Futuyma and Wasserman (1980) showed that rarer species of oaks are more likely to be defoliated than are commoner species, an observation at odds with Root's resource concentration hypothesis.

In my own research, collards grown in a weedy background accumulated more predatory Coccinellidae, Chrysopidae, and Syrphidae but fewer phytophagous insects than did collards located where weeds were tilled (Horn, 1981). The opposite effect was noted for parasitic Hymenoptera, which were both more numerous and more diverse on weedless collards (Horn, 1984, 1986). Overall species diversity (Shannon–Weaver H) was nearly identical for the insect fauna on collards within each cropping system. A "diversity–stability hypothesis" was either supported or refuted, depending upon which portion of the food web was under investigation.

Murdoch (1975) and May (1973, 1975) doubted that much practical use could be made of a diversity–stability relationship in pest management, even were one to exist in natural ecosystems. One reason is that natural systems have been in existence long enough for coevolution (genetic feedback) to be operative, whereas agricultural ecosystems are "ad hoc" assemblages of species in a more or less continuous state of disruption, with little chance for evolution of stable relationships. Most crop plants (in the United States are introduced species, whereas many of their pests are native, and the associations have existed for less than 400 years. Furthermore, "natural" plant chemistry has been significantly modified in response to breeding of agricultural plants for other purposes (high yield, palatability to humans, etc.). Comparison of agricultural with natural, evolved ecosystems is therefore imprecise and potentially misleading. Murdoch (1975) suggested that comparison of agricultural systems to laboratory models and simulations might be more appropriate. May (1975) showed that complex multi-species simulations were inherently unstable (very sensitive to perturbation). Miller (1982) suggested that systems of long-term stability may exhibit lower diversity than do systems showing intermediate levels of disturbance. Most agricultural crops are (or were derived from) species typical of early stages in plant succession, as are most of their insect pests. Agricultural operations such as soil preparation and harvesting periodically disrupt these habitats and reset the system to time $= 0$.

Usually, there is insufficient time for ecological equilibrium to become established in such ephemeral environments, and therefore there is no opportunity for coevolution of more stable relationships. Mixed cropping may be most effective against specialized herbivores, and attractiveness of the crop may vary according to the herbivores' methods of host location.

7

BROAD-SPECTRUM CHEMICAL INSECTICIDES

Insecticides are by far the most widely used single technique to reduce densities of insect pests, for reasons discussed in this chapter and the next. The advantages of using insecticides are self-evident. Chief among them is that the proper insecticide, properly applied at the proper time, nearly always causes swift death to most insects in the treated ecosystem. Insecticides are therefore especially useful in emergencies, when a rapid reduction in insect population density is necessary to prevent serious economic loss or to reduce a danger to public health. Insecticides present a wide range of choices differing in toxicity, stability, ease of application, and expense. Insecticide application is well-suited to mechanized agriculture; with modern machinery, large amounts of insecticide can be applied with minimal input of labor. Despite relatively recent increase in costs, insecticides remain cost-effective in most cases if one considers only their short-term benefit, as is customary in most economic analyses (Chapter 2). In the United States, an estimated average return on insecticides in 1960 was $2.82 per dollar invested (Pimentel, 1973), ranging to a maximum of $30 per $1 when DDT was first used against potato-infesting insects in Wisconsin (Metcalf, 1982). For all pesticides, Pimentel *et al.* (1980) estimated a return of $4 per dollar invested. Such short-term economic advantages persist at present (Pimentel & Levitan, 1986). It is therefore not surprising that annual insecticide production is increasing worldwide, exceeding 200 million kilograms in the United States alone (Metcalf, 1980).

Detailed discussion of the characteristics and uses of insecticides is a subject that would (and does) fill numerous additional volumes.

This chapter and the next review characteristics of commonly used insecticides, and the discussion then focuses on their ecological impacts and role in integrated pest management. There is sufficient difference between the effects of so-called "broad-spectrum" chemicals and those of "narrow-spectrum" (or "biorational") insecticides to justify considering each in a separate chapter.

Broad-spectrum chemicals, discussed in the present chapter, adversely affect many living creatures besides insects, and therefore most of them are properly considered to be "biocides." They enjoy widespread and intensive use, yet this use requires that great care be taken to avoid unwarranted exposure to the applicator, other persons, and nontarget organisms in the surrounding environment. Indiscriminate use of broad-spectrum insecticides has had many undesirable consequences. The prudent pest manager must be aware of the potential for acute and chronic poisoning, and for other long-term economic and ecological problems associated with insecticide use. As far as possible, the use of insecticides should be justifiable both economically and environmentally, though short-term economic gains often conflict directly with efforts to maintain environmental quality. Resistance of pests, their resurgence following insecticide use, and the development of secondary pest problems have all too often followed widespread use of broad-spectrum insecticides. Some side effects, such as hazard to nontarget organisms (humans included), are well-understood, whereas research on other effects (such as impact on soil microbes) has not been extensive.

Insecticides are the most effective means of reducing insect densities to the artificially low levels reflected in EILs for cosmetically clean produce such as fruits, vegetables, or nursery stocks. Acceptable pest densities for these may be lower than the minima that can maintain pathogens in the pest population (Kalmakoff & Miles, 1980) or below the minimum population density necessary to sustain populations of parasitoids or predators (Brown, 1978). In such cases, it may be difficult to achieve the desirable goal of successfully integrating insecticides into a framework of chemical, biological, and cultural control in an environmentally sensible pest management system. Practical examples illustrating this problem are discussed in Chapter 12.

Most insecticides exert a density-independent effect on insect populations. Application of a chemical insecticide usually controls insects below the EIL, but population density is not regulated about a mean, as is (ideally) the case with biological control (Chapter 9). Therefore, if the EIL does not change, reapplication of insecticides is usually necessary to achieve continuing low densities of pests.

CLASSIFICATION OF INSECTICIDES

A diverse array of broad-spectrum chemicals is readily available and widely used for insect management. However, each management system is unique, and the challenge of an enlightened approach to insect pest management is to choose the insecticide appropriate for the task at hand, in order to manage pest populations effectively with as little environmental disruption as possible.

For convenience here, a rating system for toxicity of insecticides is presented (Table 7-1), and insecticides are classified according to their chemical structure (Table 7-2). The discussion is brief, as more detailed information is readily available (Ware, 1978, 1983; Brown, 1978; and others). Kenaga and End (1978) is a standard reference for chemical characteristics of most major commercial and experimental insecticides.

Inorganics

Inorganic insecticides lack carbon and are usually metallic compounds or salts of copper, sulfur, arsenic, or lead. (Organic chemicals, whether or not they are actually derived from animals or plants, contain carbon). Some inorganic chemicals were used as insecticides over 2,000 years ago, and they were applied with increasing frequency commencing in the 19th century and into the earlier part of the 20th century. Inorganics, along with botanical insecticides (discussed below), were instrumental in advancing a philosophy of insect control based primarily on chemical treatment long before introduction of DDT and other synthetic organic compounds during the 1940s (Chapter 1). Inorganics are primarily stomach poisons and therefore are effective only against insects with chewing mouth parts. Lead arsenate is perhaps the most familiar of the inorganics. Originally, it was synthesized for application against the gypsy moth and was widely used in management of orchard insects. (Codling moth and leafrollers showed early resistance.) Phytotoxicity developed in a few apple orchards and replacement trees did not grow well. Lead arsenate residues are extremely stable. In general, the metallic ions from most inorganic insecticides remain reactive for long periods, particularly in the soil, resulting in the persistence of toxicity long after initial application. This persistent activity, coupled with their overall ineffectiveness as contact poisons, has caused abandonment of inorganics in favor of modern organic compounds.

TABLE 7-1
Rating System for Toxicity of Insecticides[a]

Rat: acute oral LD_{50}[b]	
Rating No.	Quantity (mg insecticide/kg body weight)
1	$> 1,000$
2	200–1,000
3	50–200
4	10–50
5	< 10

Fish: 48-hour LC_{50}[c] for bluegill or rainbow trout	
Rating No.	Quantity (parts per million)
1	> 1.0
2	0.1–1.0
3	0.01–0.1
4	0.001–0.01
5	< 0.001

Bird: acute oral LD_{50} for pheasant or mallard	
Rating No.	Quantity (mg/kg body weight)
1	$> 1,000$
2	200–1,000
3	50–200
4	10–50
5	< 10

Honeybee: topical LD_{50}	
Rating No.	Quantity (μg/g body weight)
1	> 100
2	20–100
3	5–20
4	1–5
5	< 1

Persistence: Soil halflife	
Rating No.	Time
1	< 1 month
2	1–4 months
3	4–12 months
4	1–3 years
5	3–10 years (or more)

[a]Data from Metcalf (1982). © John Wiley & Sons, Inc. Used with permission.
[b]Lethal dose for 50% of a test population.
[c]Lethal concentration per volume of water for 50% of a test population.

TABLE 7-2

Relative Toxicity and Persistence of Some Common Insecticides[a, b]

	Toxicity					
	Rat	Fish	Bird	Honeybee	Persistence	Sum[c]
Botanicals						
Nicotine	4	4	2	5	2	17
Permethrin	2	4	2	5	2	15
Pyrethrin	2	4	1	5	1	13
Resmethrin	1	4	1	4	1	11
Chlorinated hydrocarbons						
DDT	3	4	2	2	5	16
Dicofol	2	1	2	1	4	10
Endosulfan	4	5	4	4	3	20
Lindane	3	3	2	4	4	16
Methoxychlor	1	3	1	2	2	9
Toxaphene	3	4	4	1	4	16
Organophosphates						
Azinphosmethyl	4	3	2	4	3	16
Carbophenothion	4	2	4	4	2	16
Chlorpyrifos	3	3	3	5	3	17
Demeton	5	2	5	2	2	16
Diazinon	3	2	5	4	3	17
Dichlorvos	4	1	3	5	2	15
Dimethoate	3	1	4	5	2	15
Disulfoton	5	3	5	2	3	18
Ethyl parathion	5	2	5	5	2	19
Malathion	2	2	1	4	1	10
Methomyl	4	4	3	4	2	17
Methyl parathion	4	1	5	5	2	17
Mevinphos	5	3	5	4	1	18
Naled	2	2	3	4	1	12
Oxydemeton-methyl	3	2	4	2	2	13
Phorate	5	4	5	2	3	19
Phosphamidon	4	1	5	3	2	15
Stirofos	1	4	1	4	1	11
TEPP	5	2	5	5	1	18
Trichlorfon	2	1	2	1	1	7
Carbamates						
Aldicarb	5	3	5	5	3	20
Carbaryl	2	1	1	4	1	9
Carbofuran	5	2	5	5	3	20

[a]Data from Metcalf (1982). © John Wiley & Sons, Inc. Used with permission.
[b]See Table 7-1 for explanation of ratings.
[c]High totals indicate high acute or chronic hazard and greater difficulty of integration into multifaceted insect pest management program.

Natural and Synthetic "Botanicals"

Several insecticides are (or were) derived from plant products. As noted (Chapter 6), numerous plant species possess secondary chemicals that are insecticidal. Some of these substances are readily extracted, concentrated, and formulated for application afield. Nicotine, from tobacco, is an example of such a chemical; it is toxic to insects as well as to humans and other vertebrates. Extract of tobacco was first used as an insecticide in 1690, 200 years before nicotine was isolated and identified as the actively insecticidal compound (Ware, 1978). Pyrethrins, originally extracted from East African chrysanthemums and more recently synthesized from petroleum derivatives, are widely used as household insecticides because of relatively low mammalian toxicity, combined with a rapid "knockdown" that quickly results in dead insects (or apparently dead insects). Low mammalian toxicity and quick knockdown are characteristics that assure continuing sales of household insecticides. Natural pyrethrins are unstable on exposure to light and air, and therefore lose toxicity very quickly outdoors. Synthetic pyrethroids (e.g., permethrin) are more stable in sunlight and are thus more suited to management of agricultural pests, though they are more expensive than some of the alternative chemicals. Allethrin, for instance, is manufactured by a 13-step chemical process costing 100 times that of making an equal amount of DDT. Nicotine and pyrethrins kill insects by interference with transmission of nerve impulses (Yamamoto, 1970). They are quickly inactivated by acidic environments, such as within mammalian stomachs. (Insect stomach contents are not usually acidic.)

Recently, D-limonene (extracted from citrus peels) has been marketed for management of fleas and lice on pets, a situation wherein low mammalian toxicity is especialy desirable. Other plant extracts are under consideration—notably, extract of the seeds from the Indian neem tree (*Azadicachta indica*), which contains an antifeedant effective against many phytophagous pests (Jacobson, 1982; Schmutterer, 1985).

Chlorinated Hydrocarbons (or Organochlorines)

This group includes compounds of considerable variety, whether one considers chemical structure, mode of action, toxicity, or persistence. All are organic molecules containing chlorine, and effect nerve impulse transmission, apparently by upsetting the normal ionic balance of potassium and sodium, perhaps by altering permeability of the cell membrane to these ions. (See any general biology textbook.) Inexpensive, useful, and persistent DDT (dichlorodiethyltrichlorethane) is the most notorious insecticide of this (or any) chemical group. The saga of DDT, from the 1948 Nobel Prize for its rediscovery to its 1973 suspension from all but emergency uses in the United States makes an interesting story (Perkins, 1982). DDT and its derivatives are carcinogenic in some strains of laboratory mice, and mimic estrogen in gulls, feminizing males (Fry & Toone, 1981). Methoxychlor is chemically very similar, though it is degraded rather rapidly in contrast to DDT (Table 7-2).

A group of chlorinated insecticides, the cyclodienes (e.g., aldrin, dieldrin, chlordane, heptachlor) are among the most persistent of all pesticides, especially when applied to soil, where they are adsorbed to fine particles. Their activity alters soil chemistry, increasing availability of numerous metallic ions (Cole et al., 1968), though the mechanism for this is not known. The persistence of cyclodienes is useful in controlling structural pests such as termites, but the potential for environmental contamination is sufficiently severe that their use in the United States has been greatly restricted in recent years (see below). For example, toxaphene, a mixture of over 168 different specific chlorinated compounds, was widely used in management of cotton pests until its registration was cancelled in 1983 due to its persistence in the environment, its tendency to contaminate water runoff, and its mutagenic properties.

Chlorinated hydrocarbons are all contact poisons, although they penetrate insect cuticle at differing rates. They are insoluble in water and therefore are not translocated within plants. However, they show a high affinity for fats, and are concentrated in fatty tissues of animals. Once stored, most are not readily excreted, which has led to negative impact on some nontarget organisms (see below).

Organophosphates

These chemicals are all derivatives of phosphoric acid. All are nerve poisons, interfering with acetylcholinesterase, an enzyme crucial to normal transmission of impulses across the synapse between nerve endings. Generally, organophosphates are not very persistent in the environment, and for this reason they have largely replaced chlorinated hydrocarbons in agricultural and silvicultural pest management. Some are extremely hazardous, as befits chemicals originally developed for human warfare. Ethyl parathion, phorate, systox, and TEPP are examples of such highly toxic organophosphates, and in the

United States their sale and use are restricted to applicators who are certified as having had specialized training in handling and applying hazardous pesticides. At the other end of the toxicity scale, diazinon and malathion are relatively nontoxic organophosphates suitable for use by homeowners and avocational gardeners. Most organophosphates are somewhat soluble in water and some are systemic in plants. Systemic insecticides are effective in managing insects, such as aphids, whiteflies, and scales, that feed on the poisoned phloem sap but that do not move around on the plant surface and so are unlikely to contact a lethal dose of a nonsystemic insecticide during their wanderings. Systemic insecticides are also more appropriate for integration into pest management systems in which conservation of natural enemies is essential.

Carbamates

Carbamates are derived from carbamic acid, and some (but not all) have much lower mammalian toxicity than do most organophosphates (Table 7-2). Carbaryl, a carbamate, is useful in controlling a variety of insects in situations in which exposure to humans (or other vertebrates) is likely (e.g., on fresh produce near the time of harvest, or on pets or poultry). Several carbamates are relatively nonpersistent and safe to use near harvest (though a few persist for over a year in soil, given the proper cirumstances). Carbamates' mode of action is similar to that of organophosphates: chemical interference with nerve impulse transmission. Most are extremely toxic to Hymenoptera, and care must be taken to avoid accidental exposure to foraging bees or parasitic wasps. Several carbamates are translocated in plants and are useful systemic insecticides.

Oils

Oils are inert organic chemicals that are formulated for application as insecticides. They kill by coating the insect itself or the surface of calm water containing air-breathing insect (usually mosquito) larvae. Oils clog insects' spiracles, resulting in suffocation. They are a class of insecticides to which resistance has yet to develop, and such resistance is not expected to develop. However, the suffocating characteristic of oils also clogs stomata on leaves, and produces other phytotoxic effects, so that their use against agricultural or ornamental plant pests is largely limited to application onto dormant plants—for instance, to kill eggs of scale insects, aphids, and spider mites.

Synergists

Synergists are compounds that increase toxicity when added to an insecticide. Piperonyl butoxide is a synergist that, when added to pyrethrins, enhances the knockdown effect by inhibiting enzymatic oxidation of pyrethrins and thereby increasing the speed at which the nervous system is poisoned. Piperonyl butoxide also has slight anti-juvenile hormone activity (Staal, 1986). Synergism is sometimes an undesirable side effect when insecticides are mixed with certain fungicides, as in management of orchard pests. Phytotoxicity may ensue, and pest managers who mix insecticides with fungicides or herbicides must beware of potential phytotoxic synergism if they wish to remain in the fruit business rather than selling firewood. Specific information on such potential is included on the product label. Alkaline water hydrolizes some insecticides and renders them ineffective; addition of a buffer rectifies this problem. Some herbicides enhance the toxicity of insecticides. For example, atrazine synergizes carbofuran, DDT, and parathion by increasing speed of penetration through the insect cuticle (Lichtenstein *et al.*, 1979).

FORMULATION

In addition to its chemical structure, an insecticide's physical formulation is an important factor determining both its effectiveness in managing insect pests and the severity of undesirable side effects. Insecticides are formulated as gases, liquids, or solids, often with additives to enhance such characteristics as ease of application and persistence. The discussion below highlights some of the major formulations; usually, there is great variety among formulations from which a pest manager can choose.

Gaseous Formulations

Fumigants are insecticides applied either in gaseous form, or as a liquid or solid that quickly vaporizes under ambient temperatures. Use of fumigants is limited to enclosed spaces, and to soil where the gas is trapped beneath the soil surface. Methyl bromide and hydrogen cyanide (usually applied as potassium or sodium cyanide) are two widely used through quite hazardous fumigants normally applied to food-storage facilities, warehouses, and, occasionally, to dwellings under the supervision of trained personnel. Paradichlorobenzene is manufactured in a crystalline formulation widely used by homeowners for control of clothes moths and carpet beetles. Dichlorvos-impregnated wax strips are useful in controlling flying insects within enclosed spaces.

Liquid Formulations

Most insecticides are formulated for application as a liquid spray, usually as a wet or dry concentrate mixed with water. Perhaps the most familar of these is emulsifiable (or emulsible) concentrate (EC), in which an insecticide added to water is dispersed into suspension, just as liquid soap or detergent emulsifies in water. Emulsifiable concentrates are easily stored and applied (assuming readily available water), although there is danger in handling some of the more toxic compounds without proper protective equipment, because the insecticide is concentrated. Moreover, some emulsifiable concentrates cause slight phytotoxicity especially during hot, dry weather. Sometimes, it is the emulsifying agent, rather than the insecticide, that causes the wilting and discoloration commonly termed "burning."

Wettable powders (WPs) are packaged and sold dry, or formulated with a wetting agent, and are mixed with water before application; the exceedingly fine particles are suspended in the water. There is a need for continuous agitation throughout application, lest the particles settle and gum up the filter or nozzles. (Most mechanical application equipment has built-in agitation.) WPs are a bit safer to handle than are ECs, as dry insecticide is less readily absorbed through skin, and small children are less likely to eat powder from a bag than to drink a liquid that looks appetizing. WP formulations are also less likely than ECs to "burn" plants. WPs are sometimes formulated as a thick paste, termed "flowable."

Ultralow volume (ULV) formulations are commercial formulations that resemble ECs in being liquid (and often odoriferous) but ULVs lack an emulsifying agent. ULV formulations consist of undiluted chemical insecticide intended for direct application with specialized equipment. It is the development of highly accurate and dependable nozzles that has permitted the widespread use of ULV. The resulting spray is exceedingly fine and thus more prone to drifting.

This problem is particularly acute when the insecticide is applied from aircraft, for an estimated 58% of the toxicant applied drifts away from the target site. ULVs are economical for aerial application, where conservation of weight is necessary. With ULV, one can treat a far greater area from an aircraft than if one had to land often because of the additional weight of more water. Moreover, as very little solvent is applied, evaporation and runoff are minimized. Limiting runoff conserves natural enemies in ground cover near crops such as fruit trees (Hull & Beers, 1985). Hand-held ULV sprayers are useful in semiarid regions, such as the Sahel in Africa, where availability of water is intermittent (Kumar, 1984).

Aerosols are finely divided, tiny liquid particles that are suspended in the air and are most useful in enclosed areas (such as indoors) where air movement is limited. Aerosols function rather like fumigants. There is little residual effect and therefore little if any impact against insects that do not fly. Despite their limitations, and the fact that they are the most expensive formulation (by volume), aerosols are the formulation of choice for homeowners who, for the sake of convenience, are willing to pay up to 30 times the price paid by a farmer for an equivalent amount of the same insecticide in EC or WP formulation (Ware, 1978).

Solid Formulations

Dust formulations are finely divided particles mixed with an inert carrier (usually, fine clay) and broadcast onto plants where coverage is needed on both upper and lower surfaces. Coverage is often uneven, and (as in aerial application of ULV) over 80% of the dust may drift away from the site of application, even under nearly calm winds (Metcalf, 1980). Thus, dusts are inefficient in terms of the proportion of insecticide that actually reaches the target insect.

Granules are larger particles, 60 microns or more in diameter, that are often incorporated into soil at planting. The insecticide is released slowly and, therefore, granules have great persistence. For instance, granules of diazinon applied at planting provide season-long protection against cabbage maggots. Drift is minimal, and highly toxic materials (such as aldicarb) may be handled with relative safety. The greatest immediate hazard in the use of granules is the possibility of accidental mixing of insecticide with livestock feed. This is easily avoided by proper safety precautions, but it does happen.

Slow release of insecticide can also be accomplished via encapsulation. Tiny plastic or nylon spheres are impregnated with insecticides to increase the residual activity of the compound; and to reduce the acute hazard to the aplicator and his/her associates. There is some evidence that bees collect encapsulated insecticide because of the similarity of the spheres to pollen, and some cases of hive poisoning have resulted. Nylon and some plastics are virtually indestructible by normal biotic degradation, and the potential effect of innumerable microscopic spheres on natural ecosystems, water filters, etc., is a subject for lively speculation, though there is little evidence of any measurable impact. Encapsulated methyl parathion may be useful when integrated with biological control; it does not adversely affect predatory mites, for instance (Hull & Beers, 1985).

DEVELOPMENT AND TESTING

Most often, corporations in the insecticide business are interested in developing chemicals that kill a broad spectrum of pests so that the potential market is maximized, with as little hazard to the crop, the applicator, and nontarget organisms, as possible. In the United States, the specific steps in insecticide testing are prescribed by federal and state laws, most notably the Federal Insecticide, Fungicide, and Rodenticide Act of 1972 (FIFRA). The provisions of FIFRA are executed by the U.S. Environmental Protection Agency (USEPA). Before 1970 insecticide registration was under the jurisdiction of the U.S. Department of Agriculture (USDA), and the shift in responsibility generated some confusion and ill will that has waned only gradually.

The initial step in insecticide development is to measure toxicity against that of standard compounds. Usually, cockroaches, flies, rats, and goldfish are exposed to different concentrations of the compound, and an LD_{50} (lethal dose for 50% of a test population) is determined. The test animals represent species that are readily accessible and easily reared, though their physiology, like that of all organisms, is unique. A specific effect observed in rabbits or rats might not necessarily occur among humans or wildlife. That is the chance we take, sometimes to the detriment of nontarget species. For instance, DDT and related chlorinated hydrocarbons were tested extensively against gallinaceous birds (chickens, quail, etc.), yet the physiological effects of eggshell thinning observed in birds of prey came as a surprise to many scientists. The ideal insecticide displays a low LD_{50} for a wide array of insects, yet is relatively harmless to vertebrates, and is easily formulated, packaged, and stored.

Testing of insecticides in the field is undertaken by both the corporation and an independent testing laboratory, such as a State

Agricultural Experiment Station. Insecticides are applied under field conditions, often using a cooperating farmer's crop, in replicated trials against both an untreated control and a standard insecticide application recommended for use against specific pests in that geographic region. Insect populations are sampled in treated and untreated control plots at intervals after spraying to determine the residual activity of the insecticide. Turpin and York (1981), among others, have cautioned against misinterpretation of the results of such field trials. Often, researchers choose areas with unusually high population densities of a pest—densities that are not representative of the insect's population under most circumstances. Furthermore, there may be a tendency toward selectively reporting data that have the greatest statistical significance. Plot size is an important consideration; results of small-plot studies may not be directly applicable to management of huge fields.

Throughout testing, samples of foilage (or water, or whatever medium to which the insecticide has been applied) are taken periodically and examined for insecticide residue. In this way the disappearance rate of the chemical may be obtained. Insecticide use presents a paradox: it is necessary to apply sufficient amounts to kill pests, yet it is simultaneously desirable to have as little as possible present on a crop at harvest, especially when a crop treated with a broad-spectrum biocide is intended as food for humans or livestock. Phytotoxic effects and, especially in recent years, effects of the insecticide on nontarget organisms are also noted. This has not always been required, so that relatively few studies include effects on nontarget organisms at the outset of the registration process. Crops are screened for both the insecticide itself and (potentially more noxious) metabolites. For example, DDT is converted to more stable DDE, equally hazardous to vertebrates. Residue analysis for DDT alone indicates very little "insecticide," though DDE may be present in quantity.

Once all the data are taken and collated, they are submitted to the USEPA with an application for registration of the label. Along with granting the registration, and after consultation with the Food and Drug Administration (of the U.S. Public Health Service), the USEPA sets a "tolerance" or "action level," the maximum allowable amount of insecticide on a commodity at harvest. Rates of application and waiting periods until harvest take the action level into account. If it is exceeded, the produce may be removed from distribution and destroyed, and the pesticide label is likely to be suspended or at least cancelled, pending investigation of the incident. Under a notice of cancellation, the insecticide may continue to be sold and applied, pending additional research and legal appeals. Suspension is more severe, because all uses must immediately cease. A suspension order may be lifted after investigation, if it has been determined that excessive residues are due to unusual and very local circumstances.

The USEPA may also, at its option, specify that the chemical be for "restricted use," for sale only to applicators who have been certified by the appropriate agency. This agency is usually the state Department of Agriculture. An insecticide may be registered for restricted use in all formulations or only a few, usually depending on the degree of acute hazard to humans. For instance, as of 1985, carbofuran was restricted in formulations of 40% or more active insecticide, but greater dilutions were not restricted, whereas parathion was restricted in *all* formulations. In most states, pesticide applicators obtain certification by attending training sessions in safe application and recognition of pests and their damage, followed by written examination.

Finally, the USEPA periodically issues "Rebuttable Presumption Against Reregistration" (RPAR) when there is suspicion that the risks of using a specific compound outweigh the benefits. This again sends scientists and administrators scrambling for data on the extent of the insecticide's use, and the availability and expense of appropriate substitutes. A joint federal–state program, the National Pesticide Impact Assessment Program (NPIAP), coordinates some of this effort.

The total cost of developing an entirely new insecticide, including compliance with FIFRA, may take up to 10 years and $28 to $45 million (Dover & Croft, 1986). This has dissuaded the development of many new chemicals, in comparison to the situation prevailing from the late 1940s through the early 1960s, when registration cost a mere million or so dollars and new compounds were more readily available.

In the United States, each state has its own laws that are more or less consonant with federal regulations governing insecticide sales and use. Overseas, the laws are another matter. Some nations (e.g., Sweden and Great Britain) have pesticide laws even more restrictive than those of the United States, whereas others, (e.g., Italy), have fewer restrictions (Berardi, 1983). The tradeoffs between economic development and environmental protection raise some interesting and enormously complex issues, particularly in developing nations (Chapter 12).

SIDE EFFECTS OF INSECTICIDE APPLICATION

For applying insecticides there are hundreds of techniques, limited only by the ingenuity of agricultural engineers. It is not the purpose of this book to review application technology in great detail. It is sufficient to note that the farm machinery industry has developed a seemingly infinite array of application equipment, despite which the amount of insecticide reaching the target insects remains low, from less than 1% to around 5% in most cases, though some estimates of applied compound reaching the target insect pest run as low as 10^{-7} (Pimentel & Levitan, 1986). The bulk of the insecticide goes somewhere else, potentially poisoning nontarget species, drifting out of the vicinity, adsorbing to soil, and so on. Broad-spectrum insecticides kill insect pests, but environmental side effects may be considerable. These are inevitable consequences of crudely applying biocidal chemicals of broad impact to complex ecosystems. An ecological approach to pest management must be designed to minimize undesirable side effects, such as destruction of natural enemies, resistance, resurgence, secondary pests, and food-chain complications.

Insecticides that kill arthropods (and often other organisms) indiscriminately are, of course, likely to upset biological control. Interference with the action of natural enemies is well-documented. For instance, Ehler and Endicott (1984) sampled populations of olive scale both within and outside the quarantine (malathion-treated) zone sprayed for Mediterranean fruit fly following the 1981 infestation in California. Where malathion was sprayed heavily and repeatedly, there followed outbreaks of scales that had been under successful biological control by parasitoids (mostly *Metaphycus* spp.) for many years. The classical case of biological control, that of the cottony-cushion scale by the vedelia beetle on California citrus (Chapter 9) also provides an example of interference by insecticides. Application of DDT and later malathion for scale control was sufficiently lethal to the beetles that the scales resumed outbreak populations (DeBach, 1947). Heinrichs *et al.* (1982) showed that application of decamethrin and methyl parathion resulted in resurgence of brown planthoppers on rice after destruction of predators. The problem was exacerbated by high rates of fertilization and monoculture of susceptible rice varieties. Madsen and Madsen (1982) found populations of all predators to be higher in unsprayed apple orchards, though so were numbers of codling moth, the key pest of apples. Insecticide treatments do not inevitably result in reduced populations of predators. For example, Suttman and Barrett (1979) noted increased spider populations after application of carbaryl to oats.

Removal of parasitoids and predators is one component of a general reduction in diversity that often follows the application of a broad-spectrum insecticide. Soon after the introduction of DDT it was noted that a general decline in arthropod variety followed use of DDT in forested ecosystems (Hoffman & Merkel, 1948; Hoffman *et al.*, 1949). Parasitic Hymenoptera remained at densities reduced 74% to 91% for up to 6 weeks after exposure. Pimentel (1961b) found a smaller food web on insecticide-treated collards, and higher numbers of phytophagous specialists such as flea beetles (*Phyllotreta* spp.). Suttman and Barrett (1979) noted that reduction in species diversity was especially pronounced in monoculture (of oats) when compared with an old-field ecosystem. Additionally, it is speculated that, over time, repeated application of broad-spectrum insecticides may select in favor of *r*-strategists (Dondale, 1972) in much the same way as greater *r* dispersal ability in planthoppers seems to be correlated with physical instability of habitat (Chapter 4).

Addition of any chemical to the soil ecosystem may cause significant changes in the kinds and numbers of soil microorganisms (Brady, 1982). It is well-known that bacteria and fungi are major factors in the decomposition of organic matter, but rather little is known of the specific pathways of microbial decomposition and effects of pesticides thereon. Some bacteria utilize insecticides and herbicides for nutrients, breaking them down chemically and rendering them nontoxic. Regular, repeated aplications may indeed select in favor of such microbes, reducing the efficacy of pesticides applied to soil. Several soil insecticides formerly effective against corn rootworms in the midwestern United States have recently become locally ineffective, though laboratory studies reveal no appreciable resistance in the insect populations.

Application of an insecticide can result in resurgence of a pest by removal of its own natural enemies, as noted above, or can remove the former natural controls from a different arthropod, causing it to exceed its EIL and become a secondary pest. Prior to introduction of DDT and organophosphates, the European red mite was not a pest on apples. Low densities occurred, but the mites generally were held in check by predators. After the advent of contact poisons, predators were nearly eliminated, and populations of European red mite increased unfettered; specific miticides had to be added to the apple spray program, at much greater expense to the growers. A similar scenario has developed on many other crops. Cotton, for instance, added *Heliothis* spp. (bollworms) to its pest complex after sprays for boll weevil eliminated a complex of parasitoid species that had been effective in keeping *Heliothis* under control (Chapter 12). This situation has been compounded by development of insecticide resistance within many populations of *Heliothis*.

Insecticides also may influence plant growth and yield, though the mechanisms for these impacts are poorly known. Methomyl and permethrin depress photosynthesis by up to 80% in lettuce seedlings (Sances *et al.*, 1981). Dusts applied at a high dose may reduce photosynthesis simply by screening leaves from the sun. Pesticide use may stimulate growth of the pest itself; Jones and Parrella (1984) found increased survivorship and fecundity of citrus mites when the mites were fed malathion- or permethrin-treated foliage.

Hymenoptera are unusually susceptible to most broad-spectrum insecticides, and steps must be taken to assure the protection of parasitic wasps and foraging bees when insecticides are used. (In the United States, an insecticide applicator is legally liable for damage to honeybees.) There is considerable variation in susceptibility. For example, Plowright and Rodd (1980) found that bumblebees (*Bombus* spp.) were sensitive to fenitrothion but not to aminocarb, whereas solitary bees were killed in equal proportion by each. Honeybees are generally more tolerant of insecticides than are parasitic Hymenoptera (Mullin & Croft, 1985). Powell *et al.* (1985) found that parasitoids (Aphidiidae) of aphids were much more susceptible to pirimicarb than were the corresponding predators.

Chlorinated hydrocarbons at low concentrations have been especially harmful to some fish and bird species owing to the tendency of these chemicals to become concentrated in fatty tissues, and to disrupt normal calcium transport (Schreiber, 1980). DDE, the toxic metabolite of DDT, was an early suspect in eggshell thinning in pelicans (Blus *et al.*, 1971), and declines of bald eagles, ospreys, and peregrines in the United States have been generally attributed to widespread use of insecticides. Evidence in favor of this is the partial recovery of some of these birds, following the ban on most DDT use in the United States (Grier, 1982; Risebrough, 1986), though levels of DDE have remained high in other species. Chlorinated hydrocarbons are still in widespread use in many areas of the world, so this problem is by no means solved.

Resistance

Of all the problems that have arisen from the use of broad-spectrum insecticides, none has more serious and far-reaching implications than the development of resistance. The first documented case of insecticide resistance was that of citrus scales to cyanide in the early 20th century (Melander, 1914). Until the advent of synthetic organic insecticides, only 12 cases of resistance were known. Since the late 1940s, well over 400 species of insects (and lesser numbers of species of plant pathogens and weeds) have evolved resistance to pesticides, and the list seems to be growing exponentially (Dover & Croft, 1986). Pimentel *et al.* (1978) estimated that reduced efficacy due to insecticide resistance cost $130 million, a figure probably doubled by now. Most cases of resistance have been among pests of public health importance, but many common and widespread agricultural pests have become resistant, as examples throughout this book attest. (Georghiou & Saito, 1983, is a useful reference on the status of research on resistance.)

There is no *a priori* reason why insects cannot become resistant as well to narrow-spectrum insecticides (Chapter 8), and, indeed, some have done so. However, resistance to broad-spectrum compounds is much more serious and troublesome, simply because of the pervasiveness of these chemicals and the fact that they are indiscriminate, which means that a species for which the insecticide was unintended can also develop resistance. Examples include European red mite on apples (resistant to chemicals applied against codling moth), *Heliothis* spp. (resisting insecticides against the boll weevil), and mosquitoes in Californian irrigation canals (resisting organophosphates applied to surrounding cotton fields; see Luck *et al.*, 1977).

The mechanism of resistance is a straightforward process of natural selection, and the evolution of resistance represents an outstanding though fortuitous example of evolutionary change occurring in the field. Genetic variation in any insect population makes it likely that some individuals possess genes controlling traits rendering them resistant: detoxification enzymes, avoidance behavior, and so on. Regular and intensive insecticide application exerts a strong selective force favoring the survival of those individuals possessing genes for resistance, and these genes are then transmitted to offspring. It does not take long (18 generations or so) for a majority of the population to manifest resistance (Georghiou & Taylor, 1977). The speed with which resistance evolves is generally a function of (1) the proportion of the entire population exposed, (2) the frequency of insecticide application, and (3) the reproductive rate (r) of the population (A. W. A. Brown, 1968). The rate of immigration and the mating system are also important factors. Follett *et al.* (1985) showed that resistance lasted longest in populations of pear psylla that had had the greatest proportion of the population exposed. Mathematical models show that the amount and frequency of insecticide application are directly related to the onset of resistance (Tabashnik & Croft, 1982). Inbreeding populations are likely to develop (local) resistance, but the spread of the resistance will be slower than in a freely interbreeding panmictic and widely dispersing population (Horn & Wadleigh, 1987). Resistance spreads most quickly when the trait is genetically recessive, because susceptible heterozygotes are subject to very high mortality (Georghiou & Taylor, 1977).

The actual mechanism of resistance differs among different species. In many instances, a specific metabolic pathway detoxifies organic chemicals. In other cases, insects, including species of wide dietary habits, possess mixed function oxidases (MFOs), enzymes that catalyze degradation of a variety of compounds, including insecticides, that the insect may encounter. Initially, enzyme systems such as MFOs may have evolved in response to selection by exposure to secondary plant compounds (Mullin *et al.*, 1982). Resistance may be behavioral, in which case insects that avoid contact with a lethal dose are selected.

The ability of insects (and other pests) to evolve resistance to pesticides probably dooms pesticide-based management techniques in the long run. Though it is risky to speculate in a textbook, I predict less dependence on insecticides within 50 years simply because many chemicals will no longer be effective. Dover and Croft (1984, 1986) suggested that resistance itself should be managed along with insect pest populations. They proposed that the extent of resistance be monitored (which to date has not been done in any systematic fashion), that the risk of resistance be adequately assessed (by factoring into the cost curves determining EIL; see Chapter 2), and that insecticide marketing and use be regulated so that the excesses enhancing the evolution of resistance are reduced as much as possible. Tabashnik (1986) presents an integration of population modeling with resistance management. Managing resistance necessitates fundamental changes in both pest management techniques and regulation of the pesticide industry (Brown & Brogdon, 1987).

USING INSECTICIDES IN ECOLOGICAL PEST MANAGEMENT

Despite the pervasive use of broad-spectrum insecticides and their usefulness in obtaining large numbers of dead insects quickly, the percentage of insect-caused losses in crops has remained approximately constant (about 25%) throughout the 20th century (Metcalf, 1980). Resistance alone is moving insect pest management from a "protective" stance of spraying on a calendar basis whether or not pests are present, to an "as-needed" basis, wherein an EIL is established and pest density is monitored (Whalon & Croft, 1984). It is the goal of modern pest management to increase efficiency, and to reach this goal the frequency and intensity of insecticide usage must be reduced. (However, in a few cases, insecticide usage has *increased* under integrated management; the potato leafhopper, discussed in Chapter 12, is an example.) "Insecticide management" involves the following modifications in strategy.

Preventive to As-Needed

The central role of insecticide use in effective pest management is to apply insecticides on an as-needed basis, as opposed to implementing preventive "rain of death" programs. It is critical to generate a meaningful index of insect infestation—using techniques such as sequential sampling (Chapter 3) and market analysis (Chapter 2)—and then to use the most efficient and effective application technique available. More accurate estimation of insect density aids in adjusting the timing of application. Dosages can often be adjusted as well; Newsom *et al.* (1980) showd that the amount of insecticide applied to soybeans could be reduced to half the initially recommended dosage with no loss of yield. A critical need in insect pest management is greater efficiency of application for the insecticides already available, rather than development of novel compounds.

The onset of resistance can be mitigated, or at least postponed, by shifting to a chemical of a different chemical group. Pimentel and Burgess (1985) demonstrated this on houseflies in the laboratory, though the advantage is likely to be short-lived, and in practice requires cooperation among the growers in a region. Application of mixtures should generally be avoided, as they enhance the development of resistance to both compounds at once (Metcalf, 1980). Gould (1984) developed a model showing that repellency could be used to advantage; addition of a repellent to insecticides might reduce the proportion of the population exposed to a lethal dose and thus slow the onset of resistance while continuing to protect crops.

Generality to Selectivity

Most insecticides in widespread use are neurotoxins, and their mode of action in mammals is almost identical to that in insects. Pest management systems should strive to use insecticides that pose less hazard to nontarget organisms. Metcalf (1982) devised a "pest management rating" whereby the toxicity spectrum and persistence of an insecticide were ranked one to five, with higher numbers being less desirable, owing to hazard or environmental disruption (Table 7-2). The lower the rating, the more appropriate is an insecticide as a candidate for integration into a management scheme. Hull and Beers (1985) presented a similar table for insecticides and miticides used in orchard management (Table 7-3); such information needs to be developed for all major crops in which a variety of pests and beneficial organisms is active.

TABLE 7-3

Pesticide Efficiency Ratings for Apple Pests and Natural Enemies in Pennsylvania Apple Orchards, 1984[a]

	Pests[b]						Natural enemies[c]	
	CM	EM	GA	LB	PC	WH	SP	AF
Azinphosmethyl	2	—	3	3	2	3	1	1
BT[d]	3	—	4	4	4	4	0	0
Chlorpyrifos	1	—	2	2	2	4	1	1
Dicofol	—	2	—	—	—	—	1	2
Fenvalerate	1	—	2	1	2	1	3	3
Methomyl	3	—	2	1	3	1	2	2
Methyl parathion (encapsulated)	2	—	4	2	2	4	1	1
Oxamyl	—	2	3	2	—	3	1	2
Parathion	2	—	2	2	3	3	1	—
Permethrin	1	—	2	1	2	1	1.5	3
Phosalone	2	3	2	2	2	2	1	2
Phosmet	2	—	3	3	2	4	1	1

[a]Modified from Hull and Beers (1985). © Academic Press, Inc. Used with permission.
[b]CM = codling moth, EM = European red mite, GA = green apple aphid, LB = Lygus bugs, PC = plum curculio, WH = white apple leafhopper. Pesticide effectiveness: 1 = excellent, 2 = good, 3 = fair, 4 = poor, — = not rated (probably ineffective).
[c]SP = *Stethorus punctum* (lady beetle), AF = *Amblyseiulus fallacis* (predatory mite). Pesticide rating: 0 = nontoxic, 1 = slight toxicity, 2 = moderate toxicity, 3 = highly toxic, — = data unavailable.
[d]See Chapter 8.

Chapter 8 discusses narrow-spectrum insecticides; a greater variety of selective compounds needs to be developed. Herbivores and carnivores often differ nutritionally and biochemically, and these differences can be exploited. For instance, phytophagous insects have higher levels of epoxide hydrolases, and insecticides penetrate their cuticle at different rates than for carnivores (Mullin & Croft, 1985). Prestwich et al. (1983) found that the Mexican bean beetle dealkylated 29-fluorophytosterol to the highly toxic fluoroacetate, whereas carnivorous Coccinellidae did not. Evaluation of insecticide effects on biocontrol agents should become a standard practice.

Chemical versus Integrated Control

Integration of chemical with other control techniques is an ideal for which to strive. For example, Guthrie *et al.* (1959) advocated treating only the tops of tobacco plants for control of hornworms and bud-worms. The sprays killed healthy caterpillars while conserving parasi-toids that inhabited lower portions of the plants, and less insecticide was necessary because of the smaller area covered. Prepupae and pupae of natural enemies survive insecticide application more readily than do other stages (because the level of exposure is low). For example, early application of carbofuran against the alfalfa weevil reduces mor-tality of the larval parasitoid *Bathyplectes curculionis* while it is in pupal diapause (Chapter 12). Additional examples of integrating chemical and biological controls are discussed in Chapter 12.

There is no question but that broad-spectrum insecticides will continue to enjoy a dominant role in pest management in the near future. The critical need is to manage these chemicals more effectively, to make the most efficient use of their advantages by integrating them into systems of total crop management. Whether this can be done under prevailing economic constraints remains to be seen. It is more likely that the recommendations above will be adopted in public health programs and in regional, cooperative (and government-sup-ported) ventures than in the private sector. Corporations generally strive to maximize profits, and, clearly, in the insecticide business this translates into maximizing insecticide usage, rather than optimizing it. Two thirds of U.S. crop acreage is planted to corn and cotton, and two thirds of all insecticide use is on these two crops (Pimentel *et al.*, 1980). What happens to patterns of insecticide usage on corn and cotton might be an appropriate barometer by which to gauge overall trends in insecticide application. An encouraging trend is the recent reduction in the amount and frequency of insecticide application, along with increased efficiency of coverage, on both crops (Chap-ter 12).

8

NARROW-SPECTRUM
BIORATIONAL MANAGEMENT

As discussed in the preceding chapter, many insecticides are hazardous to nontarget organisms, including humans, livestock, and wildlife. This suggests an advantage to using narrow-spectrum or "biorational" insecticides that are effective against target pests while leaving other animals, and also plants, virtually unharmed. The distinction between broad-spectrum and narrow-spectrum compounds is arbitrary, for seldom is any narrow-spectrum insecticide lethal to only a single insect species, while some broad-spectrum compounds (e.g., methoxychlor) are far less toxic to humans than are aspirin, caffeine, or some patent medicines. The operational definition used here is to consider narrow-spectrum insecticides to be those that are generally nontoxic to vertebrates and plants. These include entomopathogens, insect growth regulators, and some attractants. Other attractants and repellents are not even insecticides, because they do not directly kill insects. For convenience, they are discussed in this chapter.

The chief advantages to the use of narrow-spectrum insecticides address the major acute and environmental hazards of broad-spectrum insecticides. Table 8-1 shows characteristics of some narrow-spectrum insecticides (to compare with Table 7-2). Because narrow-spectrum insecticides generally affect only insects (indeed, often a narrow range of insects), hazards to humans, livestock, or wildlife due to accidental exposure are minimal, and consequently less disruption of ecosystem functions is likely. Application of biorational insecticides is generally compatible with other agricultural management activities, for most can be applied using the same equipment as for broad-spectrum insecticides, herbicides, or fungicides.

TABLE 8-1
Relative Toxicity and Persistence of Narrow-Spectrum Insecticides[a, b]

	Toxicity				Persistence	Total
	Rat	Fish	Bird	Honeybee		
Bacillus thuringiensis	1	1	1	1	1	5
Diflubenzuron	1	1	1	1	4	8
Methoprene	1	1	1	2	2	7

[a]See Table 7-1 for explanation of rating system.
[b]Compare with Table 7-2.

On the other hand, biorational insecticides have been developed for only a limited range of pests. The available narrow-spectrum insecticides are often more expensive than comparable broad-spectrum chemicals, because manufacture is usually more complex and the potential market is more limited. There is understandable reluctance on the part of manufacturers to develop narrow-spectrum techniques when the potential for profits is limited. In the United States, infusion of federal and state funds has assisted in the development of some narrow-spectrum insecticides, but the economic paradox remains a thorny issue in a capitalistic economic system, in which corporate policy is dictated largely by profits. Due to economic considerations, the prospects for development of a wider array of biorational compounds remain uncertain unless there is failure of conventional broad-spectrum chemicals as resistance becomes more widespread or as public outcry increases against broad-spectrum biocides.

Of course, it is certainly possible—indeed, likely—that resistance to many biorational insecticides will evolve. Resistance of mosquitoes to the juvenile hormone analogue methoprene was demonstrated in the laboratory (Brown & Brown, 1974), and limited resistance to toxins of *Bacillus thuringiensis* has been shown among field populations of Lepidoptera such as the Indian meal moth (McGaughey, 1985). There has been a tendency to regard each new advance in insect management as a panacea solving all previous problems (see Chapter 1), and in some circles of current thought, biorational insecticides are touted as the ultimate solution to all insect pest management problems. The lesson from the history of insect management is that no single technology is likely to be ideal, at least for long.

MICROBIAL INSECTICIDES

Over 2,000 pathogenic diseases have been recognized in insects and related terrestrial arthropods (Cantwell, 1974; Poinar & Thomas, 1984), and this is probably a small fraction of the total in existence. Among insect pathogens there is great variation in virulence. Some are highly lethal and contagious, whereas others are nearly always endemic and apparently have little or no adverse effect on their hosts. Fungi and viruses seem to be most virulent and are responsible for most naturally occurring epizootics (Ignoffo, 1985). There is great variability in the potential of pathogens for use in pest management, due to technical difficulties as well as biological characteristics of the microbes themselves. To date the greatest successes (and the greatest hope for the immediate future) have resulted from use of bacteria and viruses. Fungi and protozoa also cause disease in insects, and there have been and will continue to be some success in their use as well. (Nematodes are sometimes considered to be pathogens; here they are discussed in Chapter 9.) The use of pathogens for insect management conjures images of biological warfare, and perhaps of Hollywood's late-show mutants that kill all human life. In fact, most entomopathogens used in pest management are quite specific to insects and are very different from microbes that cause pathology in humans. Insect pathogens have been an unavoidable minor additive to our diets for millenia (Heimpel et al., 1973), and our physiological processes are quite well-adjusted to them.

Currently, nine microbial insecticides are registered for use in the United States (Table 8-2), and another 20 were under development as of 1983 (Falcon, 1985).

Bacteria

Bacteria of several genera have been isolated and identified as causes of disease in insects, and two kinds have gained particularly widespread use. Both are spore-forming *Bacillus* spp. Spore formation is an advantage in formulation and application because it allows long-term storage of the bacteria in latent form capable of activation when applied against pests.

Bacillus popillae (including the closely related *B. lentimorbus*) was the first microbial insecticide registered for marketing in the United States. It infects larvae of several species of Scarabaeidae, and is useful in managing species (especially the Japanese beetle, *Popillia japonica*) that feed on roots of grasses. Spores are applied to turf or

TABLE 8-2

Registered Microbial Insecticides and Their Effective Dates of Registration in the United States[a]

Pathogen	Host	Date
Bacillus popillae (and *B. lentimorbus*)	Larval Scarabaeidae	1948
Bacillus thuringiensis (BT)	Larval Lepidoptera	1961
Heliothis NPV[b]	Cotton bollworm	1975
Douglas-fir tussock moth NPV	Douglas-fir tussock moth	1976
Gypsy moth NPV	Gypsy moth	1978
Nosema locustae	Acrididae	1980
Hirsutella thompsonii	Citrus rust mite	1981
Bacillus thuringiensis israelensis (BTI)	Mosquito and blackfly larvae	1981
Pine sawfly baculovirus	Pine sawfly (*Neodiprion sertifer*)	1983

[a]Falcon (1985).
[b]NPV = nuclear polyhedrosis virus.

pasture and watered-in, or a preparation is applied to the soil when new turf is installed. The spores germinate after ingestion, and in 7 to 10 days each infected larva contains millions of bacteria that convert its internal organs to a milky fluid (whence the colloquial name "milky disease"). The larva bursts, spreading its contents, with the spores of *B. popillae*, into the surrounding soil, where the spores may remain viable for many years. *B. popillae* propagates only in living larvae of Scarabaeidae, for which artificial diets have proven elusive. Mass-production of the microbe in living larvae is expensive in comparison with production costs of granular chemical insecticides (such as diazinon), which are used more frequently for control of white grubs in the lawn- and turf-care industry.

Bacillus thuringiensis (colloquially called BT) is a pathogen lethal to most species of phytophagous Lepidoptera. Over 800 different isolates are known. Ingested spores are activated in the insect's midgut, then enter the hemocoele where the bacteria replicate and release several crystalline proteins, some of which (notably the delta-endotoxin) are lethal. The lethal effect occurs faster and at lower dosage in early-instar larvae. Moreover, the toxin is synergized by a number of inert compounds (e.g., sodium benzoate and some salts of calcium, copper, or zinc) that may become useful as additives to BT (Salama, et al., 1985). BT is manufactured by fermentation, so its production is not as expensive as that of *B. popillae*; yet the cost of BT still exceeds that of comparable broad-spectrum chemicals. There are several strains of BT, some of which are toxic to aquatic larvae of Diptera, especially blackflies (Simuliidae) and mosquitoes. *B. thuringiensis israelensis* (BTI) is highly infectious in mosquito larvae and is very effective in the field against species that have been notoriously difficult to manage since they evolved resistance to chemical insecticides (De-Barjac, 1978). BTI is also effective against blackfly larvae and is used regularly in the northeastern United States and Canada (Back *et al.*, 1985).

Genetic engineering holds some promise for improvement of BT; in the laboratory, plasmids have been used to transfer genetic material between strains (Faulkner & Boucias, 1985). Approval of procedures for field-testing such genetically altered microbes is being finalized by the USEPA, although there is some concern (overflowing into heated debate) about the implications of release of engineered bacteria beyond the sterile, quarantined conditions of the laboratory (Fuxa, 1987).

Viruses

Of the numerous viruses identified in insects, the most important in management (so far) are nuclear polyhedrosis viruses (NPVs) and granulosis viruses (GVs). About 40% of identified insect viruses are NPVs (Maramorosch & Sherman, 1985). NPVs affect numerous species of Lepidoptera and members of a few other orders (especially sawflies, Tenthredinidae). The latent virus is ingested, and after activation it replicates in the midgut, resulting in inclusion bodies, or polyhedra (after which these viruses are named). Behavior may be affected; infected larvae often crawl upward before dying, and this results in more efficient spread of the virus. GVs affect Lepidoptera only, and the infection is concentrated in the fat body. NPVs and GVs have been implicated in declines of outbreak populations in forest Lepidoptera (Chapter 5; Anderson & May, 1980), and their use has some potential in management of forest pests at endemic levels to present outbreaks (Morris, 1980). A decided advantage to the use of insect viruses is that they can remain infective for up to 5 years in soils and then reinfest the insect host without reapplication of formulation from the outside. In Hawaii, Tanada (1968) found that NPVs and GVs maintained populations of the armyworm, *Pseudaletia unipunctata*, at levels well below their EIL. The worldwide resistance of *Heliothis* spp. to insecticides on cotton has created keen interest in use of *Heliothis* NPV in management. On a major, widespread crop, such as cotton, there is strong economic incentive for production of narrow-spectrum insecticides.

Unfortunately, at present insect viruses are reproducible only in living systems, and this adds to their expense. Research into artificial diets for mass-rearing of hosts and more efficient means of arthropod cell and tissue culture are necessary if production costs are to be reduced (Sherman, 1985). Recent advances in insect tissue culture and use of biological systems to manufacture complex organic molecules (including viruses) may revolutionize their production in quantity at affordable prices. DNA sequences are known for four NPVs, and current experimentation in genetic engineering is directed toward increasing stability and virulence (Faulkner & Boucias, 1985). It may also be possible (and desirable) to increase the range of potential hosts through genetic engineering.

Fungi and Protozoa

Fungi infest insects regularly only under conditions of high humidity; generally, 90% or higher relative humidity is necessary for spore germination (Maddox, 1982). Formulation and storage have presented difficulties in commercial production of fungi, and thus their use in management is limited, though naturally occurring fungal infestations can be important in pest population regulation at times. An example is that of *Erynia* sp., which control the alfalfa weevil in southern Ontario (Harcourt *et al.*, 1977; Nordin *et al.*, 1983; see also Chapter 4). *Hirsutella* is a fungal pathogen that has been effective in field tests against citrus rust mites and is used on a limited basis for its management in commercial orchards (Table 8-2). *Verticillium lecani* is commercially available for control of aphids in greenhouses in Europe (Hall & Papierok, 1982). *Beauveria* spp. infest a wide variety of pest insects and are under intense scrutiny as potential agents for pest management.

Protozoa, mostly microsporidians, are important disease agents in some insect populations, as naturally occurring epidemics may be key factors in limiting these insects (Hill & Gary, 1979). *Nosema* spp. seem to have the greatest potential for practical application in insect pest management. In particular, *Nosema locustae* has shown promise in management of rangeland Acrididae, and is now registered for that purpose. Morris (1985) showed that its rate of infection was enhanced by a low dose of dimethoate (an organophosphate), apparently due to stress induced by the chemical. A potential role for insect-pathogenic protozoans may thus be in combination with a broad-spectrum insecticide applied at a lower rate.

Effects of Microbial Insecticides

The impact of a microbial insecticide on an insect population varies, and epizootics are dependent on a complex of factors besides densities of host and pathogen (Kaup & Sohi, 1985). In general, the result of a natural or anthropogenic epizootic is overcompensating density dependence, similar to the "crash" phase in the synoptic model of Southwood and Comins (Chapter 5) with a time delay (Morris, 1977). Excepting fungi (and nematodes), microbial insecticides are stomach poisons, and therefore some time passes after exposure before the victims cease feeding and (eventually) die. Usually, feeding ceases within 24 to 48 hours. Because a few days may pass before the death of affected insects, there is some insect activity after treatment and this is not always appreciated by those who expect the nearly instantaneous results typical of broad-spectrum chemicals. Results of field applications of viruses have not been as consistent as those of broad-spectrum chemicals (Podgwaite, 1985). Older instars usually require a higher dosage, so that the timing of application is important: the earlier, the better. Most pathogens are inactivated by exposure to ultraviolet (Maddox, 1982), and so have short residual activity on agricultural crops outdoors, except when applied to soil. This is especially the case with viruses, though Ramoska *et al.* (1975) reported that *Autographa* NPV was activated upon exposure to ultraviolet.

Anderson and May (1980) and others have suggested that insect viruses are largely responsible for sudden declines in outbreak populations, at least of defoliating caterpillars, and are thus a key factor in cycling of populations of insects that are prone to periodic irruption. For such insects, it is important to establish the EIL and to determine its relationship to the population density at which a disease epidemic is initiated. A critical population density of the pest is necessary to maintain a pathogen, and a chemical insecticide may (temporarily) reduce density below that critical level (Anderson, 1982). In all too many instances such a hasty decision to use a broad-spectrum insecticide maintains an insect population at a high and damaging level just short of that at which an epidemic might commence.

The impact of microbial insecticides is not invariably lethal. Nielson (1965) reported that cytoplasmic polyhedrosis virus of *Alsophila* reduced fecundity after almost all caterpillars had survived to adulthood. Sometimes, virulence increases as additional generations of a host are exposed to a virus (Veber, 1964). The infectivity of a virus may be enhanced if the insect host is under stress, which may result from infection by another microorganism—for example, a microsporidian (Smirnoff, 1972). Insect pathogens have no discernible effect on vertebrates (Buckner & Cunningham, 1972) though there is some question as to whether this is likewise true of toxins or metabolites. For instance, the delta-endotoxin of BT causes a dermal rash in humans.

Development and Application

Legally, microbial formulations are insecticides, and as such are subject to all the provisions of FIFRA, just as are chemical compounds, with one proviso: microbial insecticides are exempt from tolerance or action levels and hence may be applied to food crops (if registered for such use) until and including the day of harvest. As noted, most are expensive to culture and formulate, and there has been a reluctance to pursue registration for fear the economic rewards may not be sufficient. This is especially true for those (e.g., *Bacillus popillae* and viruses) that must be propagated in living tissues. The cost of field-testing microbial insecticides was about $150,000 in 1985, only a fraction of that for field-testing organic chemicals.

Most microbial insecticides are applied by using the same techniques as for conventional chemicals—that is, usually in liquid form through standard spray equipment. Bacterial spores and viruses are often formulated as a fine suspension in water or oil as "flowable." Additives such as spreaders and stickers are included to increase coverage and reduce runoff, and protectants may be added against ultraviolet radiation to extend residual infectivity. Some success has been reported with addition of feeding stimulants to attract larval Lepidoptera to treated surfaces (Podgwaite, 1985). The chief difficulties in development of viral insecticides have been instability and lack of standardization (Sherman, 1985).

Pathogens are also spread by natural means. As mentioned earlier, infected larvae often climb to a high point, such as a treetop, before succumbing, and this behavior enhances spread of infective pathogens when the larvae burst. Some viruses are transmitted mechanically on the ovipositors of parasitic Hymenoptera (Levin *et al.*, 1983).

Release of preinfected hosts has generally not brought as efficient control as has spraying. Inoculation of the environment along the lines of "classical" biological control (Chapter 9) may be effective under specialized circumstances. Marschall (1970) reported long-term reduction in coconut rhinoceros beetle populations in western Samoa after inoculation of rotting logs with the specific virus *Rhabdionvirus orycyes*.

The impact of a microbial insecticide is greatest when the insecticide is applied early against smaller insects, in which case an economic threshold lower than the EIL must be determined (Chapter 2) to compensate for the lag in effect. Morris (1980) suggested that microbial insecticides were perhaps best utilized as a preventive treatment rather than an "as-needed" prescription treatment to avoid incipient damage. Microbial insecticides are useful in integrated management when a broad-spectrum chemical might upset preexisting biological control of another pest. In many European greenhouses, spider mites, whiteflies, leaf miners, and aphids are controlled by a complex of parasitoids and predators that are extremely sensitive to most broadspectrum organic insecticides. Occasional outbreaks of cutworms (Noctuidae) are managed by application of BT. In my own research on aphids, I sometimes treat experimental collard plots with BT to reduce feeding by larvae of diamondback moth (*Plutella xylostella*) and cabbageworm (*Pieris rapae*).

The most appropriate use of microbial insecticides may be in soil ecosystems, where the microbes are protected from the attenuating effects of disiccation and ultraviolet radiation, and where broad-spectrum chemicals often are not as effective. Care must be taken in integrating microbials with broad-spectrum chemicals, though most chemicals are compatible with entomopathogens. However, BT is inhibited by numerous compounds. Among other combinations, methyl parathion inactivates *Heliothis* NPV, and malathion kills spores of *Nosema* (Ignoffo, 1985). (Of course, fungicides applied to crop plants or soil inhibit the action of any fungal insecticides.)

INSECT GROWTH REGULATORS

Insect growth regulators (IGRs) are chemicals that alter normal growth and development of insects (and other arthropods), rather than killing by disrupting the nervous system or cellular respiration as do the broad-spectrum biocides. IGRs are suitable for integrated pest management because their effects are usually limited to insects and related arthropods; they are generally harmless to vertebrates, mollusks, and plants. IGRs may not directly cause the death of target insects, though they invariably reduce or eliminate reproduction due to their impact on development, and the exposed population declines (especially if immigration is minimal). IGRs are sometimes called "third-generation" insecticides (the first two generations having been inorganic salts and broad-spectrum organic chemicals). IGRs have been touted as a long-term solution to the needs of insect pest management, though they are prone to developing some of the same problems as broad-spectrum chemicals, (e.g., resistance). The major sorts of IGRs and their characteristics are discussed below.

Chitin synthesis inhibitors interfere with normal production of insect cuticle. Once an insect is exposed to a chitin synthesis inhibitor, there is insufficient chitin in the new exoskeleton at moulting, and the insect fails to moult properly. Death is usually due to desiccation, starvation, or predation. Diflubenzuron is the only registered chitin synthesis inhibitor (in the United States), and is used primarily for management of gypsy moth and boll weevil. IGRs are under the same economic constraints as are other insecticides, and these two pests represent a large potential market. Diflubenzuron is a persistent insecticide in the field, and it is a broad-spectrum chemical among arthropods. However, Ables *et al.* (1975) noted minimal impact on the housefly parasitoid *Muscidifurax raptor*, and Tamaki *et al.* (1984) found that *Doryphorophaga doryphorae*, the tachinid parasitoid of the Colorado potato beetle, was unaffected by all but extreme dosages. Heynen (1985) found that parasitized *Spodoptera littoralis* tolerated diflubenzuron more successfully than did unparasitized individuals. Diflubenzuron therefore might not be disruptive to many management systems involving biological control. It shows promise in management of codling moth in apple orchards without interfering with biological control of phytophagous mites (Moffit *et al.*, 1984). Some data indicate that nitroaniline, a metabolite of diflubenzuron, is carcinogenic in high concentrations, and diflubenzuron is therefore not yet registered (in the United States) for use on crops that are consumed by humans or livestock. Nor is it registered for use in aquatic ecosystems, because of concern about impact on Crustacea.

Juvenile hormone (JH) *analogues* interfere with normal metamorphosis. In the "classical" model (Wigglesworth, 1959) juvenile hormone (secreted by the corpus allatum) prevents maturation of gonads and other adult structures in immature insects. Reduction in concentration of JH results in completion of metamorphosis; consequently, application of JH perpetuates immature stages. This strategy is obviously most practical against insects whose major damaging stage is the adult—for example, mosquitoes and other biting flies. Methoprene was the first JH analogue registered as an insecticide in the United States for use against mosquito and flea larvae. It can also be supplied to livestock as a feed additive, resulting in JH-impregnated manure as an inadequate food supply for stable, horn, and face flies. Methoprene itself is not precisely a JH analogue, but rather is a "juvigen," which stimulates activity of the corpus allatum. The most appropriate targets for application of JH analogues are species with a short life cycle and little migration to establish reinfestations. Some of the more sedentary mosquitoes (e.g., *Anopheles* spp.) and urban cockroaches have these demographic characteristics and are often resistant to conventional broad-spectrum compounds. McNeil (1975) found that methoprene was as disruptive to parasitic Hymenoptera as to host Lepidoptera, and suggested that it may not be useful in situations in which insect parasitoids and predators are expected to be effective agents of biological control.

Precocenes (Chapter 6) are also under study. Those investigated to date result in deficiency of JH in circulation, though the precise mode of action is unclear. Exposure to precocenes leads to premature or disrupted metamorphosis. Some precocenes are selective (such as fluoromevalerate, which affects only Lepidoptera), whereas others show toxic effects even in mammalian liver or kidneys (Staal, 1986).

Other IGRs are still in experimental stages, though several may eventually join the list of registered insecticides. Among these are *ecdysone analogues,* mimics of hormones that induce moulting in insects. Immature insects exposed to a high dosage of an ecdysone analogue would moult before the new cuticle was fully developed and would die in short order. Another IGR, "Prodrone," is a chemical that alters development of fire ants, resulting in an excess of males and eventually death of the colony for lack of sufficient foraging workers. It has proven effective in limited field trials to date. Hydroprene is a hormone mimic preventing maturation of cockroach eggs.

The difficulty of developing IGRs is illustrated by the saga of kinoprene. This compound reduces fecundity and is specific to Homoptera. Costs of development were high, as were manufacturing costs, and, although kinoprene was marketed for several years for control of aphids and whiteflies in greenhouses, it was never economically competitive with systemic organophosphate and carbamate insecticides used against these pests, and was eventually withdrawn from the market in the United States. This again underscores the ongoing economic problem in the development of specific and less environmentally hazardous pesticides. Financially responsible corporations generally concentrate on developing insecticides that can be widely marketed.

SEMIOCHEMICALS

Semiochemicals are organic molecules produced by animals or plants and mediating behavioral interactions between organisms (Chapter 6). They may be considered "chemical messengers" eliciting behavioral responses in the receiver (Law & Regnier, 1971). Because their activity is often highly specific, semiochemicals offer encouraging possibilities in the management of insect pests without undesirable environmental side effects. Research in insect semiochemicals is currently a furiously active field, and recent reviews by Nordlund *et al.* (1981) and Leonhardt and Beroza (1982) are already somewhat outdated but still useful references. Chemical ecology has given rise to a terminology that may bewilder the unwary. Much of the older literature used the term "pheromone" very generally and interchangeably with "semiochemical." More recently, a generally accepted terminology for chemicals that mediate behavior has emerged (Lewis, 1981), as follows:

"Pheromones" are chemicals that elicit a behavioral response in an organism of the same species. A familiar example is that of sex pheromones, chemicals that lure females to males (or vice versa) at the time of mating. Other pheromones include trail pheromones emitted by foraging ants, alarm pheromones of aphids (Bowers *et al.*, 1972), and marking pheromones deposited by parasitic Hymenoptera (Price, 1970).

"Allelochemicals" are compounds that elicit a reaction from organisms of a different species (Chapter 6) (Whittaker & Feeny, 1971). The following three sorts of interactions occur.

1. Reactions favorable to the emitter and not the receiver are elicited by "allomones" (W. L. Brown, 1968), such as the defensive chemicals of plants that repel or reduce insect attack (Chapter 6).

2. Reactions benefitting the receiver at the expense of the emitter are stimulated by "kairomones" (Brown *et al.*, 1970). Some of the same plant defensive substances that repel one species of herbivore may be attractive to another. An example is mustard oil in leaves of Cruciferae which repels general herbivores but attracts such insects as flea beetles specific to Cruciferae (Chapter 6).

3. Mutualistic interactions that benefit both species are mediated by "synomones" (Nordlund & Lewis, 1976). Floral fragrances attractive to pollinating insects are an obvious example.

Under this terminology, the same compound can function as both pheromone and allelochemical. For instance, the marking pheromone applied by one species of parasitoid may be an allomone when detected by a second species that enhances the survival prospects of its own progeny by avoiding previously parasitized hosts.

Sex Pheromones

Of all the semiochemicals in the various stages of development and use for insect management, sex pheromones enjoy the widest applicability and potential. In most cases sex attractants have been extracted from living female insects or synthesized in the laboratory. Their use extends from monitoring insect populations to use in control. Over 100 insect chemical attractants are known (Roelofs, 1981). Enthusiastic expectations for early successes in this field have been dashed by the realization that in most cases attraction does not involve a single chemical but both long-range and short-range compounds whose activity may depend on a complex and precise blend of isomers, together with visual cues. More fundamental research on insect biology and behavior is necessary for additional progress in this area.

Monitoring

The initial successes in use of sex attractant pheromones in insect pest management were in monitoring insect populations, and pheromones continue to be widely used for this purpose. Pheromones combined with sticky traps are capable of detecting insects at extremely low densities, and are therefore especially valuable in monitoring dispersal of insect pests into previously uninfested areas. For example, the westward spread of Japanese beetle has been monitored for many years with pheromone traps containing phenethyl propionate and eugenol in 7:3 concentration (McGovern et al., 1970). Disparlure is used in thousands of gypsy-moth traps throughout the midwestern and southeastern United States. Mangel et al. (1984) recommended trapping on the perimeter of an insect infestation to determine accurately the extent of dispersal from the initial focus. Pheromones are used also to monitor insect activity for proper timing of insecticide applications, though intensive study of the biology and behavior of the insect is necessary for confidence in the samples (Whalon & Croft, 1984). Apple producers in New York reduced the frequency of insecticide applications by 50% when leafroller and codling moth populations were monitored by pheromone traps (Roelofs, 1981). Riedl and Croft (1974) suggested that results from early-season trap catches of codling moths could inform growers of expected densities in the orchard and that sprays could be planned accordingly. Shepherd et al. (1985) combined pheromone trapping with survey of egg masses of Douglas-fir tussock moth to predict outbreaks a year in advance.

Removal Trapping ("Trapping Out")

Because pheromones (and other attractants) are so efficient at attracting low-density populations, it is theorized that a proper combination of an attractant plus insecticide might result in local eradication of an insect population. Widespread application of this technique is limited by high relative cost and inefficiency of trapping in some cases; for instance, traps for Mediterranean fruit fly catch an estimated 1–2% of the actual population (R. S. Miller, personal communication). Nonetheless, this method has enjoyed some success in management of fruit flies. The oriental fruit fly was eliminated from the island of Rota after inundation with traps baited with methyl eugenol and poisoned with naled (an organophosphate insecticide) (Steiner *et al.*, 1965). Similarly, the Mediterranean fruit fly was eliminated from central Cailfornia in 1981–1982 after widespread application of malathion-laced bait sprays. However, Peacock *et al.* (1981) reported no reduction in incidence of Dutch elm disease after trapping millions of the principal vector (the smaller European elm bark beetle) via pheromone traps. Apparently enough beetles survived to maintain the spread of the pathogen. The technique has been tried against gypsy moth, with greater success in low-density populations on the periphery of the insect's range (Plimmer *et al.*, 1982). Madsen and Carty (1979) reported local elimination of codling moth populations after trapping within low-density, isolated infestations.

Confusion

The so-called "confusion technique" relies on sex pheromone dispersed in the environment so as to disorient male insects and disrupt mating, reducing overall fecundity to below the level necessary to maintain a viable population. This has led to equivocal results. The greatest success to date has been the use of gossyplure against the pink bollworm, *Pectinophora gossypiella*. Gaston *et al.* (1977) reported results comparable to those achieved with broad-spectrum insecticide but involving far less disruption to nontarget insect populations. Legner and Medved (1981), by contrast, found that the pheromone was most effective when combined with a pyrethroid insecticide. Sometimes, the specific pheromone is important in achieving the desired result: Frontalin, one pheromone of the Douglas-fir beetle, induced attack of trees, whereas another pheromone, methcyclohexanone, repelled the beetles (Furniss *et al.*, 1972). Furniss *et al.* (1974) successfully used the latter to reduce attack rate of Douglas-fir beetle and other *Dendroctonus* spp. on spruce. Sower *et al.* (1983) noted a 70–80% reduction in Douglas-fir tussock moth numbers after application of sex pheromone, though there was also a decline in untreated plots. Silverstein (1981) suggested that the confusion technique is most likely to be effective against populations of relatively low density, such as on the periphery of a species' geographical range.

Repellents

Repellent chemicals are useful in special instances when other techniques are impractical. Repellents prevent damage by deterring feeding or by causing insects simply to move away. They are therefore not actually insecticides, and protocols for their testing are more akin to those for cosmetics than for insecticidal chemicals subjected to the rigors of compliance with FIFRA. Some insecticides, such as pyrethroids and DDT, also exhibit some repellency. Many plants contain repellent secondary chemicals (Chapters 6 and 10).

Generally, repellents are effective for a brief period before loss by evaporation or chemical breakdown and are therefore not well-suited to protection of plants. The most widespread use of repellents is to manage biting flies in areas where use of broad-spectrum insecticides is proscribed due to environmental impacts, or where other management techniques are ineffective. DEET (diethyl toluamide) is probably the best all-around repellent. It provides 3 to 8 hours' protection for humans against biting Diptera (Beroza, 1972), and lasts longer on clothing or netting. Pentachlorophenol is an effective termite repellent applied to wood as a fungicide (though it contains potent teratogens as trace impurities and its use has been greatly restricted).

9

BIOLOGICAL CONTROL

Biological control, often abbreviated "biocontrol," involves importation, conservation, and encouragement of parasites, parasitoids, and predators in order to reduce pest densities to below their EILs and (ideally) to maintain them there. The basis for biocontrol is that natural enemies often exert significant mortality, reducing densities of pests (Chapter 5). Empirical evidence for this approach is that exotic pests often become much more numerous in a foreign environment lacking their specific natural enemies. Of the 212 major introduced agricultural pests in North America, 139 (65%) are not considered pests in their native homelands (Calkins, 1983). Further evidence is from outbreaks of some pests after application of insecticide disrupting activities of their natural enemies (Chapter 7). The rationale of biological control is to reestablish predator–prey or host–parasitoid equilibria, mimicking the action of natural systems, as far as possible, within the context of profitable agricultural production or tolerance of pests and their activities in urban and suburban environments. Agricultural and urban ecosystems are not natural systems, and interspecific interactions usually are more tenuous than is the case in natural, coevolved systems. Equilibrium between predator and prey may be extremely difficult to establish in anthropogenic ecosystems, and some workers (e.g., Chesson & Murdoch, 1986) suggest that predator–prey equilibrium is uncommon even in "natural" ecosystems.

Biocontrol is treated here as a process distinct from "natural" control. Populations of most insect species remain permanently below any EIL, partly due (in many cases) to the suppressive activities of a preexisting complex of natural enemies. Such natural controls exist year after year with little input from pest managers. Applied biocontrol implies active intervention with biotic components of agricultural and urban ecosystems.

When it is successful at all, biological control is relatively safe,

relatively permanent, and usually economical after the initial investment. Biocontrol is most suitable against pests with a high EIL, for
rarely, if ever, does a natural enemy exploit its prey to the point of
near-extinction required by a very low EIL, such as one finds in the
vegetable industry (in the United States). A certain minimal amount of
pest activity must be tolerated; insects are not likely prospects for
biocontrol if their feeding results in cosmetic damage at very low
densities, or if their presence represents a nascent infestation when
detected in quarantine. In fact, predatory and parasitic insects and
their fragments are considered "filth" in processed foods as surely as
are phytophagous pests (Pimentel *et al.*, 1977).

Predation and parasitism are delayed density-dependent processes
(Chapter 5) and are therefore slower than are chemical insecticides in
bringing pest densities below the EIL. Clausen (1951) suggested that it
was reasonable to expect that after introduction at least 3 years would
be required in order for newly introduced biocontrol agents to bring
their prey under control, though some pests have been brought under
biological control within a single season. Others (e.g., Horn & Dowell,
1979) noted that 3 years may not be nearly enough time for widespread
establishment of a natural enemy that eventually becomes effective.

Biocontrol is more likely to be effective in a crop with a single
major pest than in one with a pest complex. The greater the variety of
pests in a complex, the more likely it is that a single management
technique directed at a single pest species will disrupt the regulation of
another species, as the appearance of secondary pests following application of a broad-spectrum insecticide may indicate (Chapter 7). There
are a few outstanding instances in which members of a diverse pest
complex have been successfully managed through biological control.
Alfalfa insects in parts of the United States (Chapter 12), and pests of
greenhouse vegetables in Europe are examples.

Biological control encompasses the importation and enhancement of exotic agents, as well as the conservation and augmentation of
preexisting, native natural enemies. Each of these areas is discussed in
turn below. Clausen (1956, 1977) catalogued successful (and failed)
efforts in biocontrol. Texts by DeBach (1964), Huffaker (1971), and
Huffaker and Messenger (1976) discuss the principles of biological
control in much greater detail than does the treatment here.

IMPORTATION

Importation of natural enemies is handled mostly by state and federal
agencies (primarily, the USDA) in the United States, and by similar
governmental agencies elsewhere in the world. (The Commonwealth

Institute of Biological Control, CIBC, maintains a worldwide network of facilities in present and former British colonies.) The approach to importation has been largely empirical. Many agricultural pests seem to become far more damaging when introduced into new locations far from their native homeland, so that the search for effective natural enemies is concentrated in the area from which the pest is presumed to have originated. Sometimes, this is not easily determined, and thorough research in insect and plant systematics and biogeography is an essential prerequisite to successful biological control. (This fact is not always reflected among the foremost priorities of funding agencies.) Ideally, natural enemies are collected from concentrations of the pest in several geographical areas to provide as broad a range of potential biological control agents as possible. This ideal is not always reached, because of inaccessibility and/or political sensitivity. For instance, the nearest wild relatives of several major agricultural crop species are located in and around the region shared among Iran, Iraq, and the USSR, an area in which exploration by foreigners is not especially encouraged. Zwolfer *et al.* (1976) suggested that natural enemies of insects closely related to the pest should also be sought, as they too may be suitable candidates for importation (see below). For practical reasons, collection of biocontrol agents often concentrates on places where the pest is most common, though Force (1972, 1974) and others have cautioned against this, because very effective yet uncommon natural enemies may keep prey densities at low levels under natural control.

Potential agents of biocontrol are reared in strictest quarantine to ascertain details of their biology—especially, host or prey spectrum—while their precise taxonomic identity is established. The USDA and the agriculture departments and experiment stations of several states maintain rearing facilities for receipt of biocontrol agents, and similar facilities are maintained by agricultural ministries in many other nations. Knowledge of host spectrum is especially crucial in selecting agents for biocontrol of weeds, for it would not do to introduce beetles to eat weeds, only to discover later that they feed with equal voracity on agricultural crops. Generalist parasitoids and predators may exert a negative impact on native nonpest species; this has generated some concern, especially in Hawaii (Howarth, 1983). Rearing in quarantine may also reveal that the pest to be controlled is not a preferred resource for the potential agent. One must then decide whether further research is justified on that biocontrol agent.

After rearing in quarantine, if an agent is deemed suitable for release, it is shipped to release locations, often throughout the geographic range of the pest, in cooperation with state and local agencies (most often, state departments of agriculture or agricultural experiment stations). Boldt and Drea (1980) detailed techniques (involving nutrition, proper handling, and so forth) for shipping living insects. Interstate shipment of living insects in the United States for any purpose requires an appropriate permit obtained in advance from USDA–APHIS.

To establish an exotic parasitoid or predator, it is best to concentrate initial releases in areas where insecticide use is not anticipated; many a release has failed due to untimely sprays. Often, initial releases are more successful when first made within a field cage containing an artificially high pest density, allowing a local increase in the population of the natural enemy. In releasing parasitoids of the alfalfa weevil in New York, I made an effort to locate fields that remained unsprayed and uncut. In these fields, parasitoid populations increased in uninhibited fashion and became a source for repeated invasions of surrounding fields that were routinely sprayed and/or harvested. The rearing and release of insecticide-resistant natural enemies are in their infancy but hold great promise in effectively integrating biological with chemical control (see below).

EVALUATION

Experimental evidence for the success of biocontrol is rather meager, though there is an abundance of circumstantial evidence. A fundamental difficulty in evaluation of biocontrol agents continues to be lack of a generally accepted definition of "success." Definitions in the literature run the gamut from simple establishment of a novel natural enemy to permanent reduction of pest numbers below any conceivable EIL. While debate rages on, discussions on evaluation of biological control remain tentative. However, biocontrol efforts following the empirical principles of colonization and evaluation described above have led to some spectacular results generally accepted as successes, the first of which was the now-classic case of control of the cottony-cushion scale by the vedelia beetle on California citrus a century ago. Caltagirone (1981) detailed additional examples of successful colonization and augmentation of natural enemies. However, the majority of biocontrol attempts have failed to reduce pest populations appreciably, and sometimes the agent itself has promptly disappeared after release (Messenger *et al.*, 1976). Hall *et al.* (1980) found that only 16% of biocontrol programs worldwide resulted in reasonably permanent reduction in pest numbers; van Lenteren (1980) placed the worldwide rate of successful establishment at about 25%, with the rate of effective control a good deal lower. These values may seem inordinately low, yet they compare favorably with the success rate of insecticide development, in which only a few of the thousands of compounds screened initially are ever successfully used in pest management programs.

Such "before-and-after" evidence, while strongly suggesting that biocontrol might be occurring, is equivocal; the pests might have declined due to other factors. Researchers on alfalfa insects in the northeastern United States annually engage in heated debate over whether the observed decline in weevil density is due directly to parasitism by introduced Hymenoptera or due equally to adjustment in harvesting schedules and more efficient use of selective insecticides.

Key factor analysis (Chapter 5) is a more sophisticated and rigorous method to garner before-and-after evidence, and by itself does not prove that parasitism or predation (or anything else) is the actual cause of observed population declines. Any analysis of field data in the absence of a controlled experiment provides insufficient proof of successful biocontrol.

Experimental proof for biocontrol must be sought through comparative study wherein otherwise (approximately) identical plots or regions with, and without, the natural enemy in question are compared. Ideally, after a natural enemy becomes established, it may be excluded from representative plots, and a significant increase in pest density follows if biocontrol had been occurring. The best instances of evidence evaluating the impact of natural enemies are those in which natural enemies have been suppressed or eliminated locally by insecticides. The resultant resurgence of the pest is strong evidence of activity of biocontrol agents (DeBach et al., 1949), though the results can be confounded if insecticide also eliminates competitors of the pest (Root & Skelsey, 1969; Horn, 1983) or simply stimulates plant growth. It may also be possible to exclude natural enemies by caging populations of the pest, though care must be taken to ascertain that an increase in pest density is not due to changes in physical environment or to host plant conditions resulting from the cage itself. Holmes et al. (1979) demonstrated significant predation by forest birds accounting for a 37% decrease in caterpillar populations outside cages in a forest understory. DeBach et al. (1951) used ants to exclude agents of biological control from selected citrus trees, and thereby demonstrated the role of predators and parasitoids in regulating scale insects below the EIL.

ATTRIBUTES OF "IDEAL" AGENTS

Much effort has been expended in debating the attributes of effective, successful biocontrol agents. Despite a century of trial-and-error experience coupled with a bourgeoning literature on predator–prey interactions and their relationship to biocontrol, it is still unclear precisely how one selects "successful" agents of biocontrol. A relatively uncommon and insignificant parasitoid of the winter moth in England was successfully established into Nova Scotia, where it quickly brought the winter moth under control (Chapter 4). Van Lenteren (1980) concluded that, in the absence or inadequacy of pertinent theory, the empirical, trial-and-error approach was the most efficient of several alternatives. Waage and Hassell (1982) disagreed, and suggested that theoretical models at least point out characteristics of parasitoids and predators for optimal biocontrol. Their theoretical approach yielded a list of attributes more or less concordant with the one given below (see also Hassell, 1978), though the question remains open. For instance, Murdoch *et al.* (1984, 1985; Chesson & Murdoch, 1986) argued that equilibrium and "stability" rarely exist in agricultural predator–prey systems, but that local extinction is more characteristic and should be sought. Washburn and Cornell (1981) demonstrated that local extinction due to parasitoids does occur in some native gall-wasp populations. Ehler (1976) suggested that, ultimately, theoretically based biocontrol can become efficient, and to that end he utilized native host–parasitoid systems to elucidate general principles for devising strategies for releases. Ehler (1982) found that in a parasitoid complex associated with a gall-making midge native to California, a single species (*Tetrastichus* sp.) was more effective alone than was a complex of four species together. If the midge were to become a pest elsewhere, it would be logical to release the efficient *Tetrastichus*, though it might not be the most abundant species to be found during initial searches for a biocontrol agent (Ehler, 1985).

From both theoretical and empirical considerations, the following characteristics of an "ideal" agent of biological control emerge.

Host (or Prey) Specificity

Most successful instances of biological control have used natural enemies that are specific to the pest in question. In part, this is desirable to assure that an introduced natural enemy will not negatively influence other organisms in the environment. Parasitic Hymenoptera especially are usually limited in their host spectrum and therefore make likely candidates. However, Pimentel (1963) pointed out that natural enemies of insects related to the pest might also make effective agents. Hokkanen and Pimentel (1984) argued that such novel predator–prey associations are to be preferred because the prey (or host) is less likely to have evolved any resistance to the natural enemy. Drooz *et al.* (1977)

used this approach and successfully controlled loopers on pine planta-
tions in Colombia by releasing *Telonomus* egg parasitoids collected
from a closely related looper species from North America. Murdoch
et al. (1985) proposed that polyphagous natural enemies were more
efficient in regulating nonequilibrial predator–prey interactions typical of agroecosystems. Goeden and Kok (1986) cautioned that this
approach is unsuitable for biocontrol of weeds.

Synchrony

The ideal natural enemy is active at the same time as the pest. I (Horn,
1971) showed asynchrony to be the major factor causing low levels of
parasitism of the alfalfa weevil by *Tetrastichus incertus* despite high
searching efficiency by the parasitoid. The undesirable outcome of
asynchrony can sometimes be ameliorated by the presence of alternative prey to aid the agent in surviving periods of host scarcity (and
vindicating Murdoch *et al.*'s support of polyphagous biocontrol
agents).

Effectiveness at Low Host/Prey Density

The objective of biocontrol is to make a common pest rare and then to
keep it rare. The most efficient searchers therefore should be the most
likely to maintain the prey (or host) at low densities. In theoretical
terms, the sigmoid (Type III) functional response (Figure 5-2) is
shifted as far leftward as possible. This justifies looking for biocontrol
agents in locations where the pest is rather rare, in the hope that
natural enemies are maintaining the pest at low density.

Reproductive Capacity Greater than Host's

This is a most helpful characteristic aiding a natural enemy's rapid
establishment and population growth. Force (1974) suggested that "*r*-
strategists" (Table 5-1) would be most effective at suppressing pests,
though their searching efficiency might be low. Particularly, chances
for controlling *r*-strategist pests biologically are enhanced by concen-
trating effort on release and establishment of *r*-strategist biocontrol
agents (Ehler & Miller, 1978).

Dispersal Ability Greater than Host's

Outstanding successes in biocontrol have been obtained against rather sedentary pests, such as scale insects and aphids. Predators and parasitoids that are more mobile than their prey are able to locate and exploit isolated infestations more quickly.

Ease of Management

Both management of the agent in culture before release and easy integration into existing crop practices must be considered (see "Compatability," below).

Climatic Similarity

In the instances in which the influence of climate has been studied, a natural enemy has been found to be most effective in a climate similar to that from which it originated. For example, Michelbacher (1940) demonstrated that the alfalfa weevil parasitoid *Bathyplectes curculionis* from western Europe was quite effective in the relatively cool and moist climate of the San Francisco Bay region of California but was ineffective in the hot, dry San Joaquin Valley to the east. Parasitoids from the climatically similar Middle East were more effective in the more arid climate.

SINGLE VERSUS MULTIPLE INTRODUCTIONS

An ongoing debate centers on whether one should search for the single "best" natural enemy, or simply release all available potential agents, provided that they pass the tests of quarantine rearing. Turnbull and Chant (1961) championed a viewpoint shared by a strident minority of biocontrol researchers: multiple introductions may result in competition, to the overall detriment of biocontrol. Ehler and Hall (1982) found an inverse relationship between success rate and the number of agents introduced and suggested that interspecific competition may be an especially important limiting factor in the initial stages of colonization. It is apparent that some competition does indeed occur (e.g., DeBach & Sundby, 1963), yet this "classic" case turns out to be a good bit more complex than at first thought (Luck *et al.*, 1982; Chapter 5).

The bulk of empirical evidence suggests that the overall impact of a complex of natural enemies exceeds that of any one enemy regardless of its efficiency in the absence of the other species (Huffaker *et al.*, 1976). Moreover, agents of biological control effective in the initial stages of colonization, when pests are usually common, may not be as efficient later, when pests are rarer (if control is successful). A single species of natural enemy is rarely equally effective throughout its geographic range. Where possible, the most effective strategy seems to be to release a sequence of agents following (as far as practicable) the *"r-K* continuum"—the most active, fecund, voracious, and readily dispersible species first, and the most efficient, longest-lived, lowest-fecundity species last (Horn & Dowell, 1979). This often occurs anyway, because practical considerations result in initial release of the most common and most easily reared species, and *r*-strategists are efficient colonizers (by definition). The issue of multiple introductions remains another open one. Practitioners of biocontrol are effectively "restructuring" parasitoid guilds (Miller, 1983; Ehler, 1985), and are doing so in anthropogenic ecosystems that may bear little functional similarity to the ecosystems in which the original plant–herbivore–parasitoid systems evolved.

AUGMENTATION

Once exotic natural enemies have been established, their activities, and those of native species as well, may be enhanced by one or more of a number of environmental manipulations.

Inundative releases of laboratory-reared agents may supplement preexisting biocontrol to suppress pest populations threatening to exceed their EIL. *Pediobius foveolatus*, a parasitoid of the Mexican bean beetle, does not overwinter successfully in most of the United States, but mass-release of laboratory-reared wasps reduced beetle populations early enough in the growing season to slow the increase in later generations (Stevens *et al.*, 1975). Mass-release of *Trichogramma* egg parasitoids is still experimental in North America but is used routinely and quite satisfactorily against cutworms and rice borers in China (Rabb *et al.*, 1976; Feng *et al.*, 1977). Barclay *et al.* (1985) modeled inundative releases and proposed a "critical inundation rate," which (at least theoretically) led to local extinction of the prey, an exception to the general notion that biocontrol agents are incapable of controlling pests at extremely low densities. Release of mites, lady beetles, and parasitoids is routine in management of glasshouse pests in Europe (Hussey & Bravenboer, 1971). Jones *et al.* (1986) found that mass-release of the parasitoid *Diglyphus intermedius* against leaf miners in greenhouses resulted in a 25% reduction in cost, compared with that of insecticide (and, of course, resistance to insecticides was avoided). Mass-release of lady beetles in outdoor gardens yields equivocal results in aphid management despite the insistence of advertising. Overall, mass-release will probably remain specialized due to its expense, the difficulties in storing living biocontrol agents, and the seasonal nature of the venture (King *et al.*, 1985).

Alternative food sources for parasitoids or predators may be provided in times of pest scarcity, so that once the pest population increases, the natural enemies are present in sufficient number and readiness to supply effective biocontrol. Adults of many species of parasitic Hymenoptera, as well as Coccinellidae and Syrphidae, increase in both activity and fecundity in the presence of nectar and pollen, and this can result in more efficient biological control when flowers are in bloom (Leius, 1967; van Emden, 1963). Also, artificial foods may be provided, an approach pioneered by Hagen *et al.* (1970) to attract lacewings and lady beetles with a food supplement; these predators remained in the fields as aphid poulations become established. The feeding of insectivorous birds by humans during winter has undoubtedly sustained higher populations of many species in many urbanized regions of North America, though their impact on insect populations has not been determined, except in a few cases (see below). Application of allelochemicals to enhance parasitoid activity in laboratory or field is under study (Greany *et al.*, 1984). Nordlund *et al.* (1985) noted increased parasitism of *Heliothis* eggs by *Trichogramma pretiosum* when tomato extract was applied to corn.

The habitat may also be augmented in favor of biocontrol agents by provision of shelter. In North Carolina Lawson *et al.* (1961) discovered that placing in and around tobacco fields wooden sheds in which *Polistes* wasps nested greatly increased the wasps' predation on tobacco hornworms and budworms. Horn (1981) and others found that the presence of vegetational cover provided protection for natural enemies of aphids in small plots.

COMPATIBILITY

Action of natural enemies may be enhanced by greater compatibility with other crop management procedures. Either biocontrol must be successfully integrated into routine crop practices or the routine of operations must be altered, if possible. Presence of additional pest species in the system enormously complicates this integration, as is attested by the plethora of insecticide-induced pest outbreaks (Chapter 7). DeBach (1947) documented an early upset in the classic balance between the vedelia beetle and the cottony-cushion scale after application of DDT, applied by growers in response to increasing consumer demand for cosmetically clean produce.

Use of selective insecticides may be possible. Navarajan Paul *et al.* (1979) found *Trichogramma* spp. to be much less susceptible to phosalone and monocrotaphos than to malathion and methyl parathion. Reduced dosages of imidan applied against codling moth in apple orchards conserved predators of European red mites (Holdsworth, 1968). Narrow-spectrum insecticides, such as microbials, can reduce one pest (e.g., cutworms) while sparing biocontrol agents on the same crop.

Timing of insecticide applications and other agricultural operations can be adjusted to conserve natural enemies. Where possible, sprays against alfalfa weevils are applied when parasitoid activity is minimal (Armbrust & Gyrisco, 1982). Harvesting and cultivating operations may either enhance or interfere with biocontrol. Leibee and Horn (1979) reported substantial mortality among parasitoids of cereal leaf beetle after simulated plowing. Reduced tillage conserves predators, though it may also result in higher densities of some phytophagous pests. Strip-harvesting of alfalfa preserves predators and parasitoids of alfalfa caterpillars and spotted alfalfa aphids; the natural enemies simply move to uncut strips, and, a few weeks later, when these are harvested, there is sufficient regrowth in the strips harvested earlier to provide nourishment and refuge to the predators and parasitoids (van den Bosch *et al.*, 1967; Chapter 12).

An intriguing and apparently practical technique is the breeding of insecticide-resistant biocontrol agents in laboratory culture and subsequent release of these afield where insecticide treatments are routinely expected. Roush and Hoy (1980) reported success in selecting predatory mites (*Metaseiulus occidentalis*) for resistance to carbaryl, then releasing them into almond orchards treated regularly for control of navel orangeworm. Carbaryl resistance was due to a single gene, whereas permethrin resistance was polygenic; both strains are now available commercially (Beckendorf & Hoy, 1985). Headley and Hoy (1986) estimated savings of $24 to $44/acre to growers who released resistant mites. Whalon *et al.* (1982) found that laboratory-reared resistant *Amblyseiulus fallacis* survived well when exposed to applications of pyrethroids in Michigan apple orchards.

Plant resistance is generally compatible with biocontrol if spraying of insecticide is thereby reduced or eliminated. However, there may be instances in which this becomes complicated; recall that Campbell and Duffey (1979) noted that the parasitoid *Hyposoter exiguae* received toxic amounts of tomatine when parasitizing *Heliothis* larvae reared on tomato (Chapter 6). Hooked trichomes on soybeans kill foraging *Trichogramma*, and, in some soybean cultivars, antibiosis (Chapter 10) interferes with the searching behavior of adult parasitoids (Vinson, 1976; Herzog & Funderburk, 1985).

Finally, the prey or host itself may interfere with successful biocontrol. Prey possesses a variety of adaptations for defense against predation and parasitism. Many hosts encapsulate eggs and first-instar larvae of their parasitoids; up to 40% of oviposition by the alfalfa weevil parasitoid *Bathyplectes curculionis* may be encapsulated (Gibson & Berberet, 1974). Muldrew (1953) documented an instance in which a major proportion of the larch sawfly population acquired the ability to encapsulate eggs of an introduced parasitoid after 20 years' exposure. Predators of aphids may trigger an alarm reaction, causing the aphids to flee and spread (Bowers *et al.*, 1972). If these aphids vector plant pathogens, as many aphids do, the activity of predators may enhance the spread of a disease by dispersing the vectors.

MAJOR BIOCONTROL AGENTS

Relatively few of the thousands of animal species that prey on insects are suitable candidates for importation or augmentation in biocontrol programs. Most predators are either too rare or too catholic in their tastes to be of great impact on a single pest species or complex. The discussion below highlights the major groups that contain members of demonstrated benefit in insect pest management. In particular, some parasitic Hymenoptera, lady beetles, lacewings, and predatory mites are routinely used in management of pests, and some are available commercially. (The Cooperative Extension Service usually has information on their availability.) Example of the use of these biocontrol agents are found throughout this book.

Parasitic Insects

Askew (1971) remains a useful general reference on the biology of parasitic insects. Most of those having value as biocontrol agents are parasitoids (having parasitic larvae and free-living adults), and most are Diptera or Hymenoptera. Among Diptera, all members of the family Tachinidae are parasitic, usually (though not always) on larval Lepidoptera. There are over 1,300 species in North America, and several are highly host-specific (Arnaud, 1978). Examples are *Doryphorophaga doryphorae*, which parasitizes the Colorado potato beetle, and *Cyzenis albicans*, the major parasitoid of the winter moth (Chapter 4).

Among Hymenoptera, the families Braconidae and Ichneumonidae, and several families within the superfamily Chalcidoidea are parasitic exclusively on insects. Each group contains many thousands of species, scores of which have been valuable in biological control (see Chapter 12 for some examples). Many parasitic Hymenoptera have a very narrow host range, which makes them ideal candidates for biocontrol. Major impediments to their more general application are that (1) they are among the most sensitive of all insects to organic insecticides, particularly organophosphates and carbamates, and (2) the stupendous number of species frustrates the limited cadre of hymenopteran systematists (parasitic Hymenoptera are simply not very well-known, especially tropical forms). Krombein *et al.* (1979) have catalogued the North American Hymenoptera.

Predaceous Insects

Nearly all insect orders contain members that prey on other insects (or on one another), and the following are of especial value in biocontrol:

COLEOPTERA

Lady beetle (Coccinellidae) adults and larvae are voracious predators of aphids and (to a lesser extent) insect and mite eggs (Hodek, 1986). *Hippodamia convergens* is probably the most important species in North America. Ground beetles (Carabidae) are important general predators, especially of caterpillars (Thiele, 1977). Most are active at the soil surface, where they prey on such insects as cutworms and armyworms, though the large green *Calosoma* spp. are arboreal and have an impact against defoliators of forest trees.

DIPTERA

Larvae of Syrphidae (flower flies, or hover flies) are efficient predators of aphids and are common enough to exert an impact in some agricultural ecosystems (Hodek, 1986). Larvae of a few Cecidomyiidae species also prey upon aphids.

HEMIPTERA

Stinkbugs (Pentatomidae), assassin bugs (Reduviidae), damsel bugs (Nabidae), and pirate bugs (Anthocoridae) all eat immature stages of phytophagous insects (especially, small caterpillars and insect eggs). Their beneficial value seems greatest in field crops (alfalfa, soybeans, cotton, etc.).

HYMENOPTERA

Wasps (mainly Vespidae and Sphecidae) provision their nests with larval Lepidoptera and flies, among other insects. When the wasps are common, they can reduce numbers of caterpillars (e.g., Lawson et al., 1961), though they can become a nuisance. Some ant species are primarily predaceous and impact plant pests (Chapter 2).

ORTHOPTERA

Mantids (Mantidae) are large and attractive predators that are presumed by many persons to be of value in biocontrol. There is no harm in holding this opinion; however, because mantids eat almost any living insect (including one another) their value in biocontrol of any particular pest is questionable.

Arachnid Predators

Predatory spider mites (Order Acarina, several families, especially Phytoseiidae) are major predators of phytophagous spider mites and insect eggs. They are crucial to the success of many integrated management systems (Hoy, 1982, 1985). Spiders (Order Araneida) are all predaceous and probably exert an impact on many pest populations, though their role has not been well-researched. Riechert and Lockley (1984) suggested that the impact of an entire spider community (rather than of any single species) is likely to be beneficial and should be encouraged.

Nematodes

Nematodes of several families have been isolated from insects. Most insect-parasitic nematodes are Mermithidae, and Poinar (1972, 1979) documented knowledge on the best-known species. Undoubtedly, many remain to be discovered. Nematodes are very common parasites of insects, though their role in insect population dynamics is unclear. Most entomogeneous nematodes are both free-living and parasitic, entering insect hosts, in which they multiply and reproduce, sterilizing and eventually killing the host. Some transmit pathogenic bacteria that kill the host via secondary infection (Kaya, 1985). Both adult and larval insects are parasitized. Several species can be mass-produced at relatively reasonable cost (Bedding, 1984), though to date only one species has found commercial field application: *Romanomermis culicivorax* (Mermithidae), whose eggs are formulated under the trade name "Skeeter Doom." The preparation is applied to aquatic ecosystems (irrigation canals, etc.) against mosquito larvae, to which it is usually lethal. The few adult mosquitoes that survive are capable of dispersing the nematode to new sites. Generally, nematodes' impact is greatest under conditions of low light intensity and high humidity, such as in soil, under bark, or in water. Wider application of nematodes in insect management awaits economic incentive as much as technological advance (Petersen, 1982).

Vertebrates

The role of vertebrates, especially birds, as agents of biological control received much attention during the early 1900s and has attracted renewed interest more recently (M. Kennedy, 1978; Dickson *et al.*, 1979). Fish, frogs and toads, lizards and snakes, birds, and small mammals consume vast numbers of potentially pestiferous insects, and probably aid in maintaining many species at "endemic" levels (Chapter 5; also, Holmes *et al.* 1979). It is well-documented that bark-foraging birds, particularly woodpeckers, contribute measurably to mortality of borers and bark beetles (e.g., Otvos & Stark, 1985).

Less emphasis has been placed upon importation and augmentation of vertebrates than on that of parasitic and predaceous insects. Most insectivorous vertebrates eat a wide variety of insects (and other small creatures), and therefore are less likely to exert significant impact on any single pest species, though they are capable of rapid functional and numerical response. Misunderstanding of vertebrates' biology may be costly. An example is the introduction of the house sparrow into North America from Europe in a misguided attempt to control cankerworms. House sparrows are primarily seed-eaters, and their preference for nesting in and near human habitations leads to increased filth and potential disease transmission. (The sparrow is a reservoir of encephalitis virus transmissible by mosquitoes).

Some success has resulted from introduction of mosquitofish

(*Gambusia* spp.) from the southeastern United States to California (Gall *et al.*, 1980) and elsewhere. The giant toad of Middle America has been introduced widely in tropical and subtropical regions, including southern Florida. Its benefit in mosquito control has yet to be demonstrated conclusively, and it is suspected to have a negative impact on native frogs and toads.

Vertebrate populations can be enhanced by provision of suitable habitat, much the same as is done in insect predator enhancement. Feeding of insectivorous birds in winter seems to lead to increased densities of some species the following spring, though their role in insect management is not readily demonstrated. Dahlsten and Copper (1979) and others have demonstrated that provision of nesting sites for chickadees (and other cavity-nesters) results in greater mortality of forest Lepidoptera. In the eastern United States, a belief persists that provision of nesting houses for purple martins (*Progne subis*) results in reduced mosquito attacks. This stimulates the sale of martin houses (which are often an attractive outdoor decoration), though critical evaluation of the martins' role (if any) in mosquito population dynamics has yet to be undertaken. (Moreover, house sparrows or starlings routinely usurp martin houses.)

CONCLUSIONS

Biocontrol is a technique that has proven its utility and that certainly warrants increased effort and emphasis in devising ecological approaches to insect pest management. It has been more successful against pests of perennial crops, especially orchards, than in annual cropping systems, and this has been attributed to the relative temporal stability of such habitats, contrasted with the seasonal disruption brought about by plowing, planting, and harvesting (Chapter 5, Figure 5-7). Also, scale insects, which represent many of the most outstanding examples of successful biocontrol, infest mostly woody plants such as fruit crops. However, there are numerous examples of successful biocontrol in annual, ephemeral situations (Huffaker & Messenger, 1976). Perhaps the success rate on orchard crops is simply a reflection of the amount of effort expended on research and development. A greater number of successes has occurred in states (and nations) investing substantial resources in the development of biocontrol; in the United States, California, Florida, and Hawaii are noteworthy examples. In order to enjoy wider application, biocontrol must become more reliable, predictable, and compatible, especially with selective pesticides (Tauber *et al.*, 1985).

CHAPTER

10

PLANT RESISTANCE

INTRODUCTION

Plant resistance embodies the application of principles of insect–plant interactions (Chapter 6) to pest management. This chapter concentrates on resistance to insects (reflecting the author's training and biases), though resistance is even more critical in management of plant pathogens, for which other practical control techniques are often not available. Nearly all crop plant species have a measurable amount of resistance to pathogens.

Because a small residual insect population is likely to remain within the resistant crop, planting of insect-resistant crop varieties is most useful in situations where the EIL is relatively high and cosmetic damage is relatively unimportant. When effective, plant resistance has the following four positive characteristics that greatly enhance its utility in pest management:

1. Plant resistance is specific, usually limited to a single key pest species or a small pest complex. As such, it is a narrow-spectrum management technique, and (like other narrow-spectrum techniques) plant resistance minimizes environmental disruption.

2. Plant resistance is cumulative; any reduction in pest density due to resistance usually results in lowered fecundity, and subsequent generations of pests are therefore likely to be smaller than they would be on susceptible varieties.

3. Resistance is persistent, and many generations of pests may pass before resistant traits are overcome, if this happens at all. This may be a function of the hectareage under cultivation and the intensity of resistance, as well as of whether the resistance is monogenic (based on a single plant gene) or polygenic. Insect biotypes that can overcome

resistance (especially if it is monogenic) might be selected. An example is the brown planthopper (Khush, 1977), discussed below.

4. Resistance is compatible with routine crop management and maintenance of environmental quality, for it requires little or no specialized changes or expensive equipment beyond the purchase of the resistant seeds.

One of the first applications of plant resistance for insect control was the grafting of American grape rootstocks onto European varieties in the late 19th century to forestall a severe outbreak of grape phylloxera (a sap-sucking homopteran closely related to aphids). The phylloxera is native to North America, whose native grape varieties are tolerant of phylloxera feeding. After accidental introduction into Europe (around 1860) the phylloxera became extremely virulent, killing millions of vines outright until, by 1885, grape (and therefore wine) production in western Europe was on the verge of annihilation. The survival of the wine industry in France, Germany, and Italy depended on successful importation of resistant American rootstocks to which European varieties were grafted (at an eventual cost of 10 billion francs). Application of this technique persists today in the wine-producing districts of Western Europe.

MECHANISMS OF RESISTANCE

Plant resistance functions by either inhibition of a pest population or recovery of plants and production of normal yield despite attack by phytophagous insects. Painter (1951) offered a classification of kinds of plant resistance, and his classification remains widely used in textbooks and discussions of pest management techniques. Painter's terminology subdivides plant resistance into nonpreference, antibiosis, and tolerance. Nonpreference refers to situations wherein an insect simply neither feeds nor oviposits (extensively) on the resistant plant, but, rather, departs to more favored varieties or species. Nonpreference is viewed from the insect's perspective; the bug either prefers or does not prefer. For example, cotton varieties lacking extrafloral nectaries reduce pink bollworm populations by about 50% through nonpreference (Wilson & Wilson, 1976). Antibiosis is the possession of a physical or chemical characteristic that exerts a negative effect on the pests' survivorship. The impaling and killing of aphids and small caterpillars by hooked trichomes on bean leaf surfaces is an example of physical antibiosis, and the action of secondary chemical substances inimical to insects (allomones; Chapter 6) is an example of chemical antibiosis. Tolerance is the ability of a plant to withstand insect damage and continue to yield at productive levels. Antibiosis and tolerance are both attributes of the resistant plants themselves.

The classification scheme of Painter has been mildly criticized for imprecision in describing the nature of insect resistance to plants. Kogan (1982) proposed an operational classification that I consider a more precise reflection of the nature of plant resistance. He separated resistance into "ecological resistance" (primarily under the control of environmental factors) and "genetic resistance." Ecological resistance includes (1) plants' escaping insect attack due to phenological differences between plant development and insect activity ("pseudoresistance," according to Painter) and (2) induced resistance, wherein optimal fertilization and irrigation stimulate a resistant response to insect attack. Expression of induced resistance is closely related to plant nutrition, especially to the proper balance among nitrogen, phosphorus, and potassium (Kogan, 1982). Phenological asynchrony (apparent resistance from transitory characteristics in potentially susceptible plants) is not considered to be true resistance, because the insect never actually contacts the plant. For example, Kogan et al. (1974) reported asynchrony between soybean development and second-generation populations of the bean leaf beetle in Illinois, resulting in a low incidence of beetle infestation. Here this is considered to be cultural control rather than plant resistance, and is treated in Chapter 11.

"Genetic resistance" (Kogan, 1982) covers situations wherein the plant possesses genetically determined traits reducing or preventing insect damage. Genetic resistance includes antixenosis, antibiosis, and tolerance. Antixenosis (Kogan & Ortman, 1978) is analogous to Painter's nonpreference, for the insect's feeding and oviposition preference is strongly deterred by some characteristic of the plant. Antixenosis may be either physical (e.g., hairy stems) or chemical (e.g., noxious taste, oviposition deterrence, or another allomone). Antixenosis is a characteristic of the plant, whereas Painter's "nonpreference" is an insect attribute.

Antibiosis is the basis for resistance in many crop plants. It results in either direct poisoning or nutritional deficiency sufficient to reduce insect damage. It encompasses all adverse physiological effects on an insect, and may vary in impact from mild to acute. A classical example (in plant breeding) is the chemical DIMBOA, contained in most varieties of field corn grown in the midwestern United States. High concentrations of DIMBOA are present in young leaves of corn, and when the larval European corn borer feeds on the leaf, DIMBOA is converted to 6-MBOA, which is antibiotic to the larvae (Klun & Brindley, 1966).

Kogan's definition of tolerance is identical to that of Painter: the ability of a plant to repair injury and grow to produce an adequate yield despite supporting an insect population at a density capable of damaging a more susceptible host. Tolerance can be both an individual and a community characteristic. Tolerance is more subject to environmental variation than is antixenosis or antibiosis. The expression of tolerance is partially a function of how well-watered and well-fertilized is the crop. An example is that of tolerance of sorghum to the greenbug (Teetes, 1980); greenbug infestations result in reduced yields only when the tolerant plants are under stress.

DEVELOPMENT OF RESISTANT VARIETIES

Development of plant resistance is beset with several difficulties. The foremost is that many mechanisms of resistance function most effectively against a small pest complex. The more pest species involved, the greater the difficulty in breeding varieties resistant to more than one of the pests at a time. For example, frego-bract cotton confers resistance to boll weevil but increases susceptibility to tarnished plant bug (Gallun & Khush, 1980).

Development of resistant varieties requires that experimental conditions be nearly akin to those afield; too often, greenhouse tests are used for convenience, and the performance of the plants may be very different (Ortman & Peter's, 1980). Laboratory studies at constant temperature or experiments in the greenhouse can produce misleading results. Low light intensity can reduce up to 75% of resistance (Tingey & Singh, 1980). DaCosta and Jones (1971) reported that a variety of cucumber resistant to cucumber beetles in the laboratory was devastated by spider mites in the field. The beetles were attracted for oviposition by the kairomone cucurbitacin, which was simultaneously an allomone conferring resistance to spider mites. When the gene for production of cucurbitacin was bred out of the plants, the beetles could no longer effectively locate the plants, but neither was there resistance to spider mites.

Evaluation of resistant varieties in the field depends on accurate methods of sampling insects and evaluation of level of resistance (Chapter 3), and on proper statistical validation. It is not always possible to locate adequate populations of pests at precisely the time and place one wishes to screen plants for resistance, and artificially increased infestations may be used (Maxwell & Jennings, 1980). Results of such screening need to be viewed with caution.

The existence of insect biotypes interferes with effective plant resistance. The evolution of biotypes is fundamentally similar to the evolution of insecticide resistance (Chapter 7): planting of a single resistant variety over a wide area may select insects for preexisting genes that confer adaptations to overcome the resistant factor. For instance, Dunn and Kempton (1972) reported that planting of resistant varieties of Brussels sprouts throughout the principal vegetable-producing region of England eventually selected in favor of a new biotype of the cabbage aphid that was able to override resistance. A 1973 outbreak of brown planthopper on rice in the Philippines prompted the release of the resistant cultivar IR-26, planted on over 1 million hectares. By 1976, the planthopper was resistant to IR-26, and yields declined (Khush, 1977). Biotypes of Hessian fly, greenbug, and rice stem borer have all been selected simply by widespread planting of resistant wheat or rice varieties; high genetic variation in these pest species has been demonstrated in the laboratory (see below).

Most documented insect pest biotypes to date have been among aphids (Kogan, 1982), although the most thoroughly understood case is that of the Hessian fly. The Hessian fly and wheat each possess four genes determining whether the fly is susceptible or resistant to the specific cultivars (Table 10-1; Gallun & Khush, 1980). Virulence in the fly is associated with recessive genes at three loci, designated k, m, and s. Resistance in wheat is controlled by five specific dominant genes (H_3, H_5, H_6, H_7, and H_8). A Hessian fly biotype homozygous for one of the virulent loci thus will be unaffected by the antibiosis coded by the corresponding gene in wheat, and can injure the crop. For example (Table 10-1), wheat cultivar "Monon" contains gene H_3 and is there-

TABLE 10-1

Relationship between Resistance of Wheat Varieties and Biotypes of the Hessian Fly in North America[a, b]

Wheat variety (and resistance gene)	Hessian fly biotype and virulence gene							
	GP None	A s	B s, m	C s, k	D s, m, k	E k, m	F k	G m, k
Turkey (none)	S	S	S	S	S	S	S	S
Seneca (H_7 and H_8)	R	S	S	S	S	R	R	S
Monon (H_3)	R	R	S	R	S	S	R	S
Knox 62 (H_6)	R	R	R	S	S	R	S	S
Abe (H_5)	R	R	R	R	R	R	R	R

[a]R = resistant, S = susceptible. GP = "Great Plains" biotype, lacking means to overcome plant resistance.
[b]From Maxwell and Jennings (1980). © John Wiley & Sons, Inc. Used with permission.

fore susceptible to attack by flies homozygous at locus m (Biotypes B, D, and G).

Evolution of biotypes is more likely to occur when resistance is monogenic. In modeling the process, Person et al. (1952) developed the "gene-for-gene relationship," which states that for every gene coding resistance in a plant there exists within the pest population a gene for virulence and that it is only a matter of time (= selection) until a pest biotype evolves to overcome the resistant plant. Polygenic resistance is independent of biotype and hence more likely to last longer afield, despite higher initial develpmental costs (Maxwell & Jennings, 1980). Biotypes are less likely to overcome polygenic resistance. Polygenic resistance has been developed against many pests simultaneously; alfalfa is an example of a crop resistant to most major pests (Chapter 12).

Three to 15 years or more may be required from initial testing until release to growers of a new variety (Kogan, 1982). Selection has therefore concentrated on annual species, though the advent of tissue culture in plant breeding may make it easier to develop resistant varieties of plants with longer generation times, such as orchard fruits or forest trees. Plant breeders are interested in increasing yields, and this objective may override attempts to develop resistance to pests. It remains to a grower's economic advantage to plant high-yielding but susceptible varieties and then to apply insecticide, rather than to plant lower-yielding resistant varieties and reap lower profits as a result.

The nature of available genetic material itself may be limiting; truly resistant plants may not exist. Recent advances in gene splicing may artificially increase the amount of genetic material available, as it may soon be possible by genetic techniques to transplant into crop plants such genes as those coding for the delta-endotoxin of BT, for example.

Wilbert (1980) suggested that moderate resistance might be more easily integrated into pest management schemes to preserve low pest densities on which parasitoids and predators could subsist. It is important that resistant varieties that do not interfere with biocontrol agents be developed; physical or chemical factors may inhibit activity of parasitoids and predators (Bergman & Tingey, 1979; also, Chapter 6) or bees (Pimentel et al., 1984).

Historically, the development of resistant varieties of crop plants has proceeded largely on an empirical, ad-hoc basis. Discovery of resistant plants may be fortuitous, as occurred with alfalfa resistant to spotted alfalfa aphid (Howe & Smith, 1957). In this vein, the conservation of potential breeding stock is most important, and access to a large source of diverse genetic material is most desirable (Ortman &

Peters, 1980). Wild ancestors of crop plants persist in scattered locations throughout the world, often in highly restricted habitats that are themselves threatened with destruction. Wild relatives of tomato have been found to contain 70 times the toxic allelochemicals as are found in cultivated varieties (Williams *et al.*, 1980). Genes encoding such compounds could perhaps be incorporated into the crop either via traditional breeding techniques or by direct gene transfer using current biotechnological means. Wild relatives of cultivated maize exist in the high valleys of Mexico and Central America, and these species are highly resistant to many viruses to which cultivated maize is susceptible. Further economic development, including intensification of agriculture and overgrazing, threatens to destroy many such areas around the world.

A thorough grounding in insect–plant interactions, along with an understanding of the actual mechanisms underlying the resistance, is helpful, though not essential, in developing resistance. This is especially important when insect biotypes become resistant to the factor conferring resistance (Kogan, 1982). Knowledge of the role of DIMBOA in corn borer resistance has been useful in the refinement of this resistance in the United States, though the original breeding for resistance proceeded without this knowledge.

APPLICATION OF RESISTANCE

Plant resistance is generally most appropriate and is sometimes the only practical technique for management of insect pests on field crops such as grains and forages, where profit is relatively low per hectare, use of insecticides is difficult to justify economically, and minimal damage to plants is tolerable. Resistance is especially suitable for crops in developing nations, where little specialized skill or additional investment is needed by growers, once resistant varieties are developed.

Plant resistance is used in two ways: as the principal control method, and as an adjunct to other techniques in an integrated management program (Painter, 1951). Management of the grape phylloxera in Europe is an example of the use of resistance as a principal control technique. In North America, successful management of Hessian fly, first brood European corn borer, and spotted alfalfa aphid each is due principally to resistant varieties (of wheat, corn, and alfalfa, respectively). Ortman and Peters (1980) suggested that resistance was best suited as an adjunct to other management techniques. When combined with other methods, plant resistance lowers pest density and thereby lengthens the time to reach the EIL, extending the interval between applications of insecticide when necessary (Adkisson & Dyck, 1980). The restlessness of boll weevil on frego-bract cotton increases its exposure to insecticide while reducing oviposition and subsequent larval infestations (Mitchell *et al.*, 1973). Slower growth and development of a pest leads to increased exposure and mortality to natural enemies (though a few plant allelochemicals have been demonstrated to interfere with parasitoid activity; see Chapter 9). Resistant plants may thus render a pest population sufficiently low to permit effective biological control by natural enemies. In terms of Southwood's synoptic model (Chapter 5), the growth rate, r, is reduced, in effect reestablishing the "natural enemy ravine" (van Emden, 1966). Plant resistance is most effective when all growers within a region cooperate, as in the area-wide planting of fast-fruiting cotton varieties to manage boll weevil in the High Plains of Texas (Chapter 12).

At the same time, "genetic monoculture" is fraught with vulnerability to pest outbreaks precisely as is "ecological" monoculture (Chapter 6). For instance, a large proportion of blight-susceptible corn in the midwestern United States succumbed to blight in 1970, and this might have been avoided by planting several different resistant varieties. Genetic heterogeneity in a crop buffers against the evolution of insect biotypes, as well as against the disaster of crop failure.

Resistance is useful against pests that are exposed in the field for a very limited time. The Hessian fly adult is extremely short-lived, whereas its eggs, larvae, and pupae are protected within the wheat stem. The fly may be exposed and vulnerable to management by insecticide only 1 week of the year, whereas resistance protects against infestation for the entire season. (The Hessian fly is routinely managed by adjustment of planting date; see Chapter 11.)

Resistance may be either enhanced or retarded by extremes in temperature, humidity, or light, as well as soil fertility and overall stress; the influence of environment on the expression of resistance varies because the mechanisms of resistance are variable. Low temperature inhibits resistance of alfalfa to aphids and sorghum to greenbugs (Tingey & Singh, 1980). Pesticides, especially herbicides, may interfere with resistance by stressing plants or, alternatively, by increasing their nutritional quality. The array of resistant plants may enhance or retard expression of resistance. Cantelo and Sanford (1984) planted resistant and susceptible varieties of cabbage, beans, and potatoes in replicated plots, and found that antixenosis for imported cabbageworm was accentuated by planting mixed stands of cabbage, whereas potato leafhopper antixenosis was enhanced in pure stands.

Plant resistance will continue to be developed as a valuable tool in ecological pest management. The outstanding successes in management of Hessian fly, wheat stem sawfly, alfalfa aphids, and European corn borer in the United States, and of pests of rice, cotton, sorghum, and sugar cane worldwide, along with the promise of new approaches to genetic manipulation, are stimulating new research in this area. There are especially attractive prospects for further development of plant resistance in vegetable and fruit crops (e.g., G. Kennedy, 1978). Maxwell and Jennings (1980) and Mayo (1980) remain useful general references on plant resistance, together with Painter's (1951) classic monograph.

CHAPTER

11

GENETIC, CULTURAL, AND PHYSICAL CONTROL; QUARANTINES

This chapter discusses several techniques that bear little relationship to one another beyond the fact that they are neither insecticidal (Chapters 7 and 8), nor biological in the classical sense (Chapter 9), nor based on plant resistance (Chapter 10). Genetic manipulations, cultural management, and/or physical control may be used in concert, with or without chemical or biological controls in integrated management systems. For convenience, these techniques are discussed in the present chapter.

GENETIC MANIPULATIONS

The use of insect genetics in insect management holds great promise for effective pest management. Genetic manipulations are usually species-specific, and therefore their development is subject to the economic constraints noted previously for narrow-spectrum insecticides and biological control, though recent advances in virology and genetic engineering may result in economical genetic management techniques. At present there are numerous technological impediments to development of genetic manipulations; to date, the most successful techniques have been the use of insecticide-resistant biocontrol agents (Chapter 9) and the release of sterile males to reduce the fecundity of pest populations. This has led to one outstanding success and several additional experiments with promising results.

Sterile-Male Technique

The principle of the sterile-male technique is simple yet elegant: Male insects sterilized in the laboratory are released into the field to mate with wild females. To ensure success the sterile males must be competitive with wild types. If the ratio of sterile to wild males is sufficiently high, and the rate of immigration is sufficiently low, the probability is high that any particular wild female will mate with a sterile male and produce no eggs, and eventually the pest population will be reduced below its "extinction threshold" (Chapter 4); its decline is then assured, at least in the geographic region surrounding the sterile-male releases (Barclay & Mackauer, 1980). Knipling (1979) suggested release ratios of 10 to 100 sterile males per wild male. Sterile-male releases are most efficient against species whose females mate only once, though, with a sufficiently high ratio of sterile to wild males, populations of even the most promiscuous species can be reduced. The sterile-male technique is obviously limited to use against species whose adult males do no damage. Biting Diptera (except muscoid flies in which both sexes bite) and most phytophagous Lepidoptera are likely candidates.

The classical (and, to date, only) widespread effective use of this technique has been against the primary screwworm fly (*Cochliomyia hominivorax*) in the southern United States and northern Mexico. The screwworm female larviposits on fresh wounds in wildlife and domestic stock (and in humans, although very rarely). The feeding larvae enlarge the affected area until infection sets in, and, if untreated, the victim dies. No economical chemical or sanitary control exists for controlling screwworms that infest animals on open range. The screwworm female mates only once and the absolute population density is low, as are rates of dispersal, rendering this insect an ideal candidate on which to practice releases of sterile males. Moreover, the EIL is low (one untreated infestation on a steer can result in a loss of several hundred dollars), and there has been strong regional political support for the expenditure of public funds on the project (as one might expect for a pest of range cattle in Texas). The screwworm is easily mass-reared in the laboratory, male pupae are smaller than—and therefore easily separated from—those of females, and a radiation dose sufficient to sterilize apparently does not otherwise affect the flies.

After preliminary success eliminating screwworm from the island of Curacao and from southern Florida, a program of releases that eventually totaled 180 million sterilized male flies weekly was initiated in states along the southwestern border of the United States. Table 11-1 shows results of this highly successful program, which to date, has

TABLE 11-1

Confirmed Cases of Primary Screwworm in the Southwestern United States during Releases of Sterile Males (Commencing 1962)[a]

Year	Cases	Year	Cases
1961	>1,000,000	1973	15,016
1962	50,850	1974	7,273
1963	7,168	1975	17,568
1964	395	1976	29,671
1965	1,062	1977	468
1966	1,898	1978	7,230
1967	872	1979	90
1968	9,877	1980	2
1969	219	1981	5
1970	153	1982	6
1971	473	1983	0
1972	95,626	1984	0
		1985	0

[a]From Richardson et al. (1982) and Krafsur (1985). © 1982, American Association for the Advancement of Science. Used with permission.

eliminated the screwworm from the United States and most of Mexico. The reversals occurring during the mid-1970s were attributed to selection for different biotypes (Bush et al., 1976) or to mild winters (Cocke, 1981). Evidence for the former comes from more recent releases concentrated in southern Mexico (near Tuxtla Gutierrez), where Richardson et al. (1982) reported enzymatic differences among several strains of C. hominivorax, and identified at least five different karyotypes. They postulated that release of laboratory-reared flies may have changed the proportional distribution of wild strains. On the other hand, Krafsur (1985) found little incontrovertible evidence that genetic changes had occurred in laboratory screwworm cultures during the 1970s.

Sibling species and biotypes may be more common among insects than is at first evident (Makela & Richardson, 1978), though this can be used to advantage (see below). In general, it is well to commence releases on the periphery of a species' range, lest there be inadvertent selection for dispersing biotypes. The screwworm program was aided by reduction of the fly population due to cold winter weather in the northern periphery of its range.

All told, the screwworm program has been a resounding success; an early estimate (Baumhover, 1966) was that over $140 million was saved for the first $10 million spent, a 14:1 benefit–cost ratio that compares very favorably with recent estimates for conventional insecticides (about 4:1, according to Pimentel *et al.*, 1980), and with much less impact on the environment. (One adverse impact is that deer have been extremely abundant in Texas since the mid-1960s, though this is not entirely due to the screwworm eradication program.)

Experimental release of sterile males has shown promise against other pests. The mosquito *Culex pipiens quinquefasciatus* was eliminated from Seahorse Key, Florida, in 10 weeks by daily release of 8,000–18,000 sterile males (Patterson *et al.*, 1970). Stable flies on St. Croix, Virgin Islands, were reduced 99.9% after release of 100,000 sterile flies daily for 18 months (Patterson *et al.*, 1981). Sterile-male releases have been successful against fruit flies on several Pacific islands, and may have helped in the eradication of the Mediterranean fruit fly from California in 1982 (though results are equivocal because malathion bait spray was also used). High-density populations may be reduced by application of insecticide before sterile-male release; such a strategem has been effective against localized infestations of codling moth in British Columbia (Proverbs *et al.*, 1982) and against German cockroaches on board ships of the United States Navy (Ross *et al.*, 1981). Takken *et al.* (1986) reported local eradication of the tsetse *Glossina palpalis* via a combination of trapping and sterile-male releases.

The chief limitation of the sterile-male technique is that irradiation necessary to sterilize may also interfere with other aspects of growth and development. Gross *et al.* (1972) reported reduced emergence of black flies after irradiation. High radiation doses caused deformation of flight muscles in tsetse flies (Dame & Schmidt, 1970). Boll weevils lost their competitive ability, and their ability to mate was only 20% of what it had been before irradiation (Villavaso, 1981). It is also essential that there be low rates of reimmigration into areas of sterile-male releases (Asmen *et al.*, 1981) or wherever any other genetic management technique is attempted.

Additional Genetic Techniques

Other genetic manipulations are mostly in developmental stages. It is theorized that chromosome translocations can be integrated into insect genomes and spread by subsequent recombination (Wallace, 1985). Delayed sterility is also possible. Each program needs to be engineered specifically to the pest involved, and there are, and will continue to be, problems in identifying mutations that reduce fitness of the pest population while remaining transmissible by the insects that carry them. There are other technical limitations, such as development of artificial diets. Nonetheless, some intriguing possibilities have shown promise experimentally and are in limited use. Laven (1967) reported local eradication of *Culex pipiens fatigans* in Burma via mass-release of males bearing cytoplasmic incompatibility. Rogoff (1980) suggested release of autogenous (nonbiting) strains of stable flies. Similarity, strains of face flies that avoid exudates from cattle exist, and mass release of these strains could reduce spread of conjunctivitis among cattle. Collins *et al.* (1986) discovered that a strain of *Anopheles gambiae* that encapsulated the malarial parasite could be propagated in the laboratory; such a strain would be useful for release in the field. Seawright *et al.* (1978) used a sex-linked propoxur resistance on the male Y-chromosome to separate males from females of *Anopheles albimanus*; once the gene was incorporated into the laboratory culture, the mosquitoes were sprayed and only males remained. This has yet to be attempted on a large scale in the field. Hoy and Knop (1981) selected pesticide-resistant predatory spider mites in the laboratory, and these were released into vineyards and apple orchards to prey upon phytophagous mites that otherwise are difficult to manage in agroecosystems that require heavy insecticide exposure due to the demand for cosmetically clean fruit. The genetic engineering of biocontrol agents holds considerable promise in development of integrated pest management (Chapter 9).

CULTURAL MANAGEMENT TECHNIQUES

A variety of agricultural practices are important in reducing or eliminating insect damage. Some of these are directed against specific pests, whereas others result in increased pest mortality as a by-product of normal cropping operations. In all cases, cultural management is most effective when used against the most vulnerable life stages, and underscores the need for a thorough understanding of pest biology and ecology in order to pinpoint key factors and other vulnerabilities. There are many agricultural practices that mitigate insect damage, a few of which are highlighted below.

Especially in temperate climates, autumn plowing and disking exposes overwintering forms of many insects to predation, freezing, and so on. In the midwestern United States, the European corn borer overwinters as a mature larva in corn stalks left in the field after harvest. Plowing of stubble exposes these larvae and results in mortality that may exceed 90%. Turning of sod exposes larval Scarabaeidae to predation, and large numbers are eaten by birds in the autumn. Destruction of crop residue by burning or burying, where feasible, is a very effective tool for management of both insect pests and plant pathogens.

Where irrigation is regularly practiced, fields can be flooded to eliminate plant pests. Irrigation generally must be carefully regulated; cotton, for example, can be severely stressed by too little water, but too much will result in overly lush plants with higher insect densities and more difficult harvesting problems (University of California, 1984).

Crop rotation is a time-honored technique for reducing insect infestations and forestalling increases of many monophagous insects. For instance, if corn and soybeans are rotated on a 2-year cycle, large populations of corn rootworms, white grubs, and wireworms are prevented from developing, though some of the "generalist" pests (e.g., seed-corn maggot and cutworms) are unaffected. Rotations among legumes, grasses, and root crops are usually especially effective, though corn planted after grass hay may develop high wireworm populations. Unfortunately, economic factors often mitigate the potential for rotation. If corn prices are significantly higher than those for soybeans, the prudent farmer will plant corn despite the potential increase in cutworms, rootworms, wireworms, and so on. Home gardeners can often eliminate some pest problems simply by leaving a particular vegetable out of the garden at intervals (Chapter 12).

Many agricultural pest outbreaks are exacerbated by monoculture, and multicropping and conservation tillage are each techniques that may aid in reducing pest infestations (Chapter 6). The intercropping of maize, beans, and squash on a single plot is a traditional technique for small-scale agriculture in tropical America. Risch (1981), Altieri *et al.* (1978), and others have documented that this multicropping scheme results in fewer insect pest outbreaks. A reduction of pest problems on urban shade trees is enhanced by planting a variety of species. The spread of Dutch elm disease among shade trees in the eastern United States stands as testimony to the hazards of monoculture. More recently, nearly monocultural stands of maples and/or honey locusts in some urban regions have become prone to a variety of insect and disease problems (Kielbaso & Kennedy, 1983). Multicropping is not well-suited to mechanized agriculture, though recent trends toward conservation tillage have essentially the same effect (Lowrance *et al.*, 1984; Gebhardt *et al.*, 1985). Selective ground covers may enhance the action of biocontrol agents in orchard crops (Chapter 9). Altieri and Whitcomb (1979) found that field-edge management to encourage growth of goldenrod and *Chenopodium* enhanced aphid numbers, which in turn produced increased populations of Coccinellidae. These predators then moved into neighboring crops. (Obviously, one should select only weeds that encouraged nonpest aphids; plants favoring the green peach aphid would not do.) Orchards with cover crops may have lowered densities of codling moth, aphids, and leafhoppers (Altieri & Schmidt, 1986). It is not yet known whether, in general, minimizing tillage enhances pest populations or those of their predators more selectively. Results may depend on specific associations; Kumar (1984) reported that *Heliothis* spp. on cotton in Africa increased when nearby plantings included corn and tomatoes, alternative hosts of *Heliothis* spp. Obviously, crop managers need to consider the agroecosystem from an expanded viewpoint that includes ecosystems surrounding the crop of interest as well as that crop itself (Altieri & Letourneau, 1982).

Sometimes, "trap crops" can be utilized to attract pests that otherwise would harm a more economically important crop (Bucher & Cheng, 1970). Planting wheat among beets has been shown to reduce wireworms (Martin & Woodcock, 1983). To control the boll weevil, cotton growers may plant a few rows 2 or 3 weeks early, and treat these with a systemic or foliar insecticide once weevils are concentrated on them. Asparagus beetles may be managed similarly by leaving a portion of the crop unharvested in spring; the beetles concentrate on the remaining mature plants in August and then may be easily destroyed by insecticide or burning. Early planting of soybeans in a small patch within a larger field concentrates bean leaf beetles, which can then be killed with insecticide. "Trap trees" have been used in Europe to attract bark beetles within commercial forests (Coulson & Witter, 1984).

Plant density affects distribution of insects (Chapter 6) and can sometimes be managed to improve insect control. Piedrahita *et al.* (1985) showed that corn rootworms were most concentrated in high-density plantings. Closely spaced peas resulted in a doubling of the population density of the beetle *Ootheca mutabilis* in Nigeria (Kumar, 1984). Conversely, the wheat-stem sawfly apparently prefers widely spaced plants, and denser stands reduces its impact.

Timing of planting (or harvest) can sometimes be adjusted to entirely avoid an insect with a very specific period of activity. An outstanding example is the case of the Hessian fly in North America. This pest has a very short adult lifespan (3–4 days, on average) and oviposition occurs over a limited span of time during early autumn. Damage to winter wheat can be avoided entirely by planting after the so-called "fly-free date" in autumn. This is most effective when all growers in a region cooperate. It is not without risk, for very wet weather in October (sometimes occurring in the midwestern United States after West Indian hurricanes) can prevent planting entirely. Similarly, early-planted peanuts avoid aphid damage in tropical Africa (Kumar, 1984). First-generation European corn borers are far less numerous on late-planted maize (Everett *et al.*, 1958), though the second generation may be larger. Usually, this is acceptable because, by virtue of larger size, maize in August is much more tolerant of the second generation of European corn borers. Short-season varieties of cotton are recommended for management of all insect pests on that crop (Chapter 12). Early-season (45-day) varieties avoid most boll weevil and bollworm populations (Adkisson & Dyck, 1980). Early-maturing (134-day) soybeans generally avoid bean leaf beetle when compared to standard 152-day varieties (Bernard, 1971).

SANITATION

Sanitation is the nonagricultural equivalent of cultural control. It involves cleanup of breeding and gathering sites, and includes draining of stagnant water for mosquito management, cutting of weeds and tall grass to reduce numbers of chiggers and ticks, and drying and spreading of manure to discourage flies. Sanitation is particularly useful on a localized level, such as in and around human habitations. Cleanup of spillage, proper insect-tight storage of dried food and woolen goods, and so on, all contribute to reduction of insect infestations.

PHYSICAL CONTROLS

A vast array of traps, barriers, and other devices has been used for centuries to manage insect pests. If one includes hand-picking, physical control antedates any other sort of management strategy. Our tree-dwelling primate relatives still control lice and ticks this way. Humankind has progressed far beyond this in innovative mayhem visited upon insect pests. Some of the barriers protecting against insects are so commonplace that we take them for granted (until we are without them). Window screening and mosquito netting are in this category. Barriers of bare ground sometimes offer protection for crops against insects such as chinch bugs, grasshoppers, and armyworms that routinely migrate into grain fields. Such barriers are far more effective if there is an insecticidal bait included.

In addition to traps involving chemical attractants (Chapter 8), light traps of various sorts make use of the propensity of many night-flying insects to orient toward a source of radiant energy. Some of these traps are combined with an electric grid in which the unsuspecting insect is electrocuted. Such traps are of some use in mitigating the nuisance of flying insects outdoors and the sight and sound of insects being "zapped" are most satisfying to the consumer, though the practical value of such devices is questionable. For instance, hungry mosquitoes often ignore ultraviolet light and orient directly toward their warm-blooded victims (Nasci et al., 1983). Ultraviolet light traps may catch large numbers of houseflies but do not appreciably reduce local infestations (Skovmand & Mourier, 1986).

Ultrasound is another physical control higly touted as a means of ridding a prescribed area of insect pests. Results of field tests of ultrasonic devices are equivocal (Rust & Reierson, 1983; Foster & Lutes, 1985) and do not justify some of the more fabulous claims of the marketers of the devices. Nonetheless, this does not dissuade purchasers of the devices. In insect management, as in other aspects of economy, some people are willing to be separated from their financial resources even in the face of objective evidence (P. T. Barnum, unpublished).

Radiation is of practical utility in managing insects infesting stored food products. Doses of 5 kilorads or more are sufficient to prevent metamorphosis of most grain pests (Brower & Scott, 1972), and this technique will probably grow in importance with the spread of insecticide resistance among pests of stored grain.

Simple heat or cold may be sufficient to kill insects of stored products. Use of heat and cold is most appropriate around the home. A temperature of $+60°C$ for 1 hour or $-20°C$ for 4 days is usually sufficient to kill grain pests. Grain-storage facilities in the upper-midwestern United States, Canada, and the Soviet Union can harness midwinter air to reduce infestations.

Fine dust or ash abrades insect cuticle sufficiently to cause death by desiccation. Though this has not been routinely applied on a commercial scale, inclusion of road dust in flour has been practiced for many centuries in portions of North Africa (Levinson & Levinson, 1985). Volcanic ash can reduce densities of both plant pests and their predators (Fye, 1983), and level of control depends on their ratios at the time of ashfall.

Finally, despite the significant 20th-century advances in chemical technology and integrated pest management, the ordinary hand-held fly swatter continues to be a most useful and effective technique applicable on an as-needed basis.

QUARANTINES

Quarantines are sometimes termed "legislative control" or "regulatory control." By strict definition, legislation by itself does not "control" insects (or anything else), but it establishes the statutory authority for governmental agencies to engage in limiting insect dispersal or in treating localized infestations of pests deemed a threat to public welfare. In the United States, certification of plants and quarantine of plant pests are under the authority of the USDA–APHIS–PPQ (Division of Plant Protection and Quarantine). Whether to issue a quarantine against a particular region or commodity in order to regulate a pest is a political decision that may bear significant expense and effort, and so is not taken lightly. The case of the gypsy moth is instructive. In North America the moth is confined primarily to the northeastern United States, with a few outlying infestations (Michigan, Oregon, California). USDA–APHIS and states on the fringe of the generally infested area deem it advisable to engage in limited quarantine activities against the moth. In Ohio, for example, a statewide network of pheromone traps is maintained by the Ohio Department of Agriculture, with partial financial support from USDA–APHIS. The traps are positioned throughout the state, though they are concentrated in trailer parks, roadside rest areas, and railway yards, places that have historically served as foci for gypsy moth dispersal. (The egg masses are often deposited on undersurfaces of vehicles that then leave the infested zone.) Consistent capture of male moths in any single locality triggers an intensive search for egg masses during the following winter, and, if eggs are located, the surrounding area (usually about 100–200 hectares) is sprayed with carbaryl (two aerial applications) the following May, and the treatment is followed by intensive pheromone trapping to ascertain that the infestation has indeed been eradicated. In this way, the gypsy moth has been "eradicated" from Ohio two dozen times during the past decade.

As a general rule, USDA–APHIS personnel consider a pest to have been eradicated if surveys are negative for three generations (Rohwer & Williamson, 1983). Even the most zealous defenders of gypsy-moth eradication programs admit that the programs are a delaying tactic, and that the gypsy moth will eventually invade the forests of Ohio and other midwestern states. However, the delay is considered worthwhile if it buys 10 more summers with leaves on the trees, and 10 more years of favorable wood growth in the extensive and commercially valuable oak forests farther west, especially those in Missouri and Arkansas.

The efficiency of federal inspections and quarantines is attested to by the interception of over 15,000 infested items annually at ports of entry into the United States. Despite this vigilance, about 1,500 alien insect species have become established in the United States since colonization by Europeans, and most have arrived during the present century. In Hawaii alone, 38% of the insect fauna have been introduced to the state, and an estimated 16 species per year colonize despite the strenuous efforts of APHIS personnel (Crooks et al., 1983). The ubiquity and accessibility of jet aircraft have reduced transit time between international ports to the point at which almost any insect (or weed or pathogen) in any life stage might survive a flight of many thousand kilometers and find favorable conditions in the new territory. Colossal amounts of baggage at ports of entry exacerbate these problems, and agricultural inspectors can be overwhelmed. For example, there is no practical way for Saudi Arabian authorities to inspect the baggage of the many millions of Islamic pilgrims going to Mecca, and many of them bring food along, which may harbor potential agricultural pests.

Early detection is essential to success of quarantine. Once established, some species spread so quickly that quarantines are futile. The cereal leaf beetle is such a case. Early after its introduction into southwestern Michigan (apparently from central Asia, though precisely how it got from there to Michigan remains a mystery), a quarantine was placed on hay and straw moving from the North Central region to the southeast. This quarantine was to little avail because the beetle, apparently aided by prevailing winds, quickly spread beyond the limits of the quarantine zone. Fortunately, the cereal leaf beetle has not become a serious pest, due in part to a successful program of biological control.

Occasionally the publicity surrounding quarantine and eradication efforts evokes some ill-advised reactions. A classic case is that of the Mediterranean fruit fly (Medfly) in California in the early 1980s. The Medfly originated in Africa and is now widespread in tropical regions. The larvae feed on a great variety of fruit and vegetable crops, and one estimate is that an infestation in the United States would ultimately result in damage of about $1 billion annually (Jackson & Lee, 1985). In the conterminous United States, the Medfly had been detected, and eradicated, seven times prior to its being detected yet again in central California in 1980. Low-intensity trapping apparently failed to detect the initial infestation, and there was a subsequent delay in action. At first, sterile-male releases (which had been successful on previous occasions) were attempted, resulting in inundation of pheromone traps intended for detection of wild flies. Ultimately, malathion-bait spray was applied from aircraft, resulting in near-panic over the potential side effects on humans of malathion-bait sprays. The California Department of Food and Agriculture initially failed to make a case in support of the sprays against some rather preposterous and emotional rhetoric perpetrated on a gullible public in a nonagricultural setting. There were indeed side effects; for instance, Ehler and Endicott (1984; Ehler *et al.*, 1984) found a 90-fold increase in gall midge densities, apparently due to destruction of parasitoids and predators. Most side effects were limited to populations of other insects, and the levels of exposure to humans were far below those having been shown to cause any ill effects in any laboratory vertebrate animal. (It is the author's opinion that breathing the atmosphere of the San Francisco Bay region during air stagnation is far more unhealthful than exposure to an environment in which malathion-bait spray has been properly applied. Moreover, the risk from the eradication of the Medfly is nowhere near that taken by most of the citizenry, each within his/her automobile, during a terror-filled sojourn on the Bayshore or Nimitz Freeways.)

International cooperation in regulating agricultural commerce is necessary, given the global nature of commerce. Through the auspices of the United Nations' Food and Agriculture Organization, 82 nations participate in an International Plant Protection Convention. Each participating nation maintains an organization, similar to APHIS–PPQ, that issues import and export certifications (Chock, 1983). Widespread containerization of cargo has made it imperative (and often easier) to initiate quarantines at the point of origin rather than at the point of importation. If current trends in international trade continue, international cooperation in enforcement of quarantines must increase.

12

INTEGRATED INSECT
PEST MANAGEMENT

The previous several chapters discussed specific tactics of pest management separately for convenience, as though each operated independently from others. Management in the real world, with its entomological, agricultural, meteorological, and economic uncertainty, dictates a multifaceted approach to insect pest management within each system. In turn, insect management is only one phase of total environmental and economic management, and a satisfactory approach to pest management must consider the entire ecosystem containing the pest in question. Many complex decisons and tradeoffs are involved, with a degree of uncertainty at every step (Chapter 2). Usually, a combination of two or more techniques provides more satisfactory and lasting management than does a single "best" technique. Dependence on biocontrol and plant resistance, with an occasional application of a selective insecticide, often gives more satisfactory control than will any single technique by itself. Such an approach is at the heart of modern integrated pest management, or IPM.

The challenge is to incorporate the complexity of multispecies ecosystems into suitable descriptive models applicable to management of crops, forests, or urban environments. Construction of realistic, descriptive, workable ecosystem models is a specialized endeavor requiring a multidisciplinary team of entomologists, plant pathologists, weed scientists, plant physiologists, economists, and systems scientists (Ruesink, 1976). The specifics of complex model-building are beyond the scope of this book, although the following discussion should serve as a useful orientation. Ruesink (1976, 1982) suggested the following four critical considerations in approaching pest management systems from a modeling standpoint.

1. Assumptions of the model must be realistic. A simulation is an approximation of the real world, and (as has been shown) simplification of assumptions for tractability may quickly render a model inaccurate as a description of reality.
2. Objectives of producers and the overall society must be considered, especially when these come into conflict. In capitalist economies, a grower's principal objectives (the "bottom line") are most often to maximize profit and reduce risk (Chapter 2), whereas the larger society may be more receptive to tradeoffs between higher prices and lower pesticide inputs into the environment.
3. Causes and effects within the modeled system must be known as completely and accurately as is feasible. Especially critical is knowledge of the biologies of both the pest and the plant and their respective responses to environmental factors.
4. Direct and indirect effects of management decisions must be known as far and as accurately as is possible.

Shoemaker (1973) proposed that a valid pest management model should fit available data, be consistent with biological reality, utilize meaningful and easily measured physical parameters, and provide predictability in the field. Getz and Gutierrez (1982) cautioned that insects and plants themselves are systems and are components of larger systems. They added that the principal pitfalls of modeling efforts have been (1) ignorance of significant aspects of pests' biologies, (2) needless complexity in models and/or specificity to one or a few sites only, and (3) failure of the model's validity over a broad array of conditions.

Coulman *et al.* (1972) suggested that the following steps be taken in constructing pest management models.

1. Determine the portion of the system to be modeled, including the life stages of the insect involved. This includes damaging stages, as well as other stages (if any) most vulnerable to management techniques.
2. Select system components that reflect primary functions—that is, identify and model key factors of greatest importance in population trends (Chapter 4); for simplicity, the fewer of these that are included, the better.
3. Convert biology to mathematics to computer programs by providing an equation for each component: input, state variables, and output.

4. Couple model components; the output of each component simply becomes the input of the next. Serial life tables, in which the eggs resulting from females of one generation become the starting cohort for the next, are an example of such a model.

Operational insect population models are sometimes expressed in terms of energy (calories of insect) rather than density (Gutierrez *et al.*, 1984). After a density-based or energetic model is developed to describe the population dynamics of the key pest, an insect model can be coupled with models for host plant, other insect pests, natural enemies, and economic models, to provide an overall simulation model describing the system. Growth of the plant is subdivided into photosynthesis, tissue growth, and carbohydrate utilization (respiration) (Stimac & O'Neil, 1985). Plant population processes are generally analogous to those of insects, especially in being temperature-dependent (Gutierrez *et al.*, 1980), though thresholds for plant development often differ from those for insect development. Physiological time (degree-days) is most often used because of the temperature dependency of insect and plant development (Chapter 3).

A modeling approach is best for managing pests that have a variable EIL and whose populations are easily sampled. Prime candidates are crops with a single, widely distributed key pest, and on which indiscriminate insectide use is economically unsound. An ancillary consideration is that a widely grown, economically important crop (like cotton in the southern United States) is more likely to provide the broad financial support necessary to the successful development of a multidisciplinary modeling effort.

Even when insect populations are highly variable, or when detailed population studies are not yet available, it is often possible to monitor the onset of an infestation and to relate this to an easily measured variable, such as temperature or precipitation. This has been successfully accomplished for several insects (alfalfa weevil, European corn borer, etc.; see Gage *et al.*, 1982, and examples later in this chapter).

Before implementation, pest management models must be validated. Welch *et al.* (1981) pointed out that the purpose of a model dictates the approach to validation. Research-oriented models (such as the spatial distributions of Chapter 3) are validated by standard statistical techniques, and precision and stability of variance are most critical to determining validity. Management (implementation) models designed for use in the "real world" should first be subjected to cost–benefit risk analysis, preferably in actual commercial situations. "Failure" of implementation models is simply suboptimal accuracy in describing actual events in the field and, consequently, in providing recommendations that are also suboptimal (or wrong). Valid generalized management models may lack the rigorous detail typical of an analogous research model, but that is acceptable as long as the management models are accurate when applied afield. Ultimately, if complex pest management models are to become useful, they must help increase (or at least sustain) agricultural profits while minimizing risk (Shoemaker, 1980). The greater the accuracy of an implementation model, the greater is its utility in managing pest populations, as long as the model is not too complex to operate.

Some systems are inherently unmodelable, due to extreme stochasticity, leading to unpredictability. Some defoliating Lepidoptera, such as bagworms (Psychidae), and some scale insect species, fall into this category. Others may not be worth modeling (as yet) for economic reasons. Among the latter are situations wherein the EIL is invariably exceeded in the absence of strenuous and disruptive inputs such as chemical poisons (though a model may be useful in refining the precision of timing insecticide application against those pests for which insecticide is routinely applied).

APPLICATION OF PEST MANAGEMENT MODELS

The most effective utilization to date for IPM models has been in computer-assisted pest management, in which a model provides the basis for prediction of a pest infestation, along with recommendations for courses of action. One of the first such "on-line" systems, PMEX, was instituted by Michigan State University in 1975, commencing with apple pest management and later expanding to include field crops and vegetables. The PMEX system takes input consisting of insect population estimates, plant development, and weather data into an internal multicomponent management model. The resulting outputs include management advisories as well as educational information, such as pesticide label updates (Croft et al., 1976).

Each year there is greater development of computerization in the field of agricultural advisories. Development of on-line pest management systems has become a very fast-moving business since the advent of personal computers. Several agricultural advisory firms, along with the Cooperative Extension Service, have developed and are developing a bewildering array of agricultural software. Most computer-assisted pest management systems are similar to PMEX in that they provide an output of management advisories after inputs of weather data, insect density estimates, and measurements of plant growth and developmental stage. Examples of such regional sytems include SCAMP (System for Computer Aided Management of Pests) in the northeastern United States, and the national AGNET system, in addition to the various state systems. Detailed, verified crop management models now exist for alfalfa, apples, citrus, cotton, and soybeans (Huffaker, 1980), with others under development and likely to hit the software stands before this book is published. These models combine with artificial intelligence and expert systems to produce advisories for growers (Lemmon, 1986; Coulson & Saunders, 1987).

Welch (1984) predicted that the use of remote terminals and the next generation of "booksize" personal computers will do much to facilitate standardization and quality of pest management data. It has been predicted that by 1990 a majority of farms in the United States will have small computers as standard equipment. Computerized databases provide speed and efficiency, increased accuracy via centralized data control, and a structured format leading to ease of error-checking (Brown et al., 1980). USDA–APHIS has instituted a national survey program, to which state IPM programs contribute. There have been some initial problems with the system; however, once the bugs are worked out, distributional data on major insect pests throughout the United States will be readily available. Similarly, the USEPA has instituted a National Pesticide Information Retrieval System (NPIRS) for rapid updating of pesticide label status. This is a welcome development in the rapidly changing world of pesticide registrations.

Undergirding all pest management decisions is the assumption that an injury level, economic or aesthetic, has been defined. A decision has been made as to how much damage can be tolerated, and how this damage is related to a measurable quantity, such as an estimate of the insect population density or some indicator of density or impact (Chapter 3), or to an easily measured predictive parameter, such as temperature. Inputs must be known with sufficient precision to generate informed decisions. The accuracy of any model, forecast, or database is obviously dependent upon the quality of input.

Despite a scientific, objective approach, one cannot know everything, and management decisions may contain an element of educated guesswork. Decision-making in IPM involves a large dose of scientific objectivity and a bit of speculation. The remainder of this chapter details the "state of the art" in a few specific management instances in order to illustrate application of principles developed in the preceding chapters. Alfalfa, vegetable, cotton, forest, and household IPM are examples; the discussion is not intended to be exhaustive. There are many comprehensive guides that detail biology and management of the pests involved in specific situations. Metcalf *et al.* (1962), Davidson and Lyon (1979), Wilson *et al.* (1980), and Pfadt (1985) are especially useful, as are some of the citations within each section to follow.

ALFALFA: SUCCESSFUL DEVELOPMENT OF MANAGEMENT MODELS

Alfalfa is a perennial, protein-rich legume grown primarily for live-stock feed worldwide in temperate climates. In the United States in 1982, 12.5 million hectares were planted to alfalfa; only corn, cotton, and soybeans exceeded alfalfa in hectareage. Most alfalfa is used either for pasture or as bulk feed for beef and dairy cattle. A lesser amount is dried and processed for use as supplemental feed for poultry and small mammals that are in laboratory cultures or kept as pets. Some is grown for seed to produce the rest of the crop; American seed production is largely limited to the western states. There are over 200 varieties of alfalfa in commercial production. When well-watered and well-fertil-ized, most varieties of alfalfa display tolerance to phytophagous in-sects. Moreover, because nearly all alfalfa is destined to be eaten by livestock or pets (which are not too finicky about insects or their fragments in food), EILs are generally rather high, and small insect populations do not normally constitute an economically important threat.

The development of pest management on alfalfa is an instructive example of the application of principles developed in previous chap-ters. Over 1,000 insect species have been found in alfalfa fields, and management depends largely on which species of the pest complex are present. In most of North America, a key pest in the alfalfa ecosystem has been the alfalfa weevil, *Hypera postica*; in the arid southwestern United States, however, it is replaced by the sibling species *Hypera bruneipennis*, the Egyptian alfalfa weevil. The alfalfa weevil, origi-nally from Europe, was introduced into the western United States around 1900. Before 1952, there was no chronic insect pest on alfalfa in the eastern states or southeastern Canada. There were periodic, minor outbreaks of pea aphids and/or meadow spittlebugs in wet years, and of grasshoppers or cutworms in dry years. This bucolic situation changed abruptly with the introduction of the weevil in 1952. The alfalfa weevil spread throughout the remaining alfalfa-producing re-gions of North America during the next two decades.

In the northern United States and Canada, the weevil overwinters as an adult, and in spring oviposits in stems of alfalfa that are more than 10 centimeters tall. Larvae tunnel into and defoliate the growing tip, then move onto outer foliage, destroying much of the nutritive value of the spring crop. Sometimes entire plants are killed, and, rarely, an entire stand (usually a first-year seeding) is destroyed.

Until 1962, damage was prevented by application of a granular cyclodiene insecticide (heptachlor) to alfalfa stubble in autumn. This provided season-long protection at low cost (around 50 cents per hectare) by killing adult weevils at the outset of the oviposition period. In its day, this was a successful preventive management program, and was typical of the indiscriminate use of highly persistent, broad-spectrum insecticide prevalent at the time.

In 1962, registration for use of heptachlor on alfalfa was cancelled after small but significant amounts of heptachlor epoxide residue were detected in milk and meat at the very time of heightened public concern over insecticide residues (Carson, 1962). Researchers then instituted an interim management program, based on organophosphate or carbamate insecticides sprayed directly onto larvae in the plant tips. When properly timed, such treatments did reduce damage to below the EIL, but were expensive, more than four dollars per hectare, plus cost of labor and equipment. Timing was critical, and a hastily adopted EIL of "75% tip damage" was difficult for growers to measure and interpret. Great care was necessary to avoid excessive insecticide residue on the crop at harvest, especially when the alfalfa was destined for feeding to dairy cattle.

Meanwhile, multifaceted programs engaged scores of investigators, who researched all aspects of the weevil's biology and weevil–plant interactions, with special efforts directed toward biological and cultural control, plant resistance, and improvement of insecticide application techniques. Twelve species of exotic parasitic Hymenoptera were established (Table 12-1), with varying success in reducing weevil densities (Dysart & Day, 1976). In the northeastern and midwestern states, the combination of *Bathyplectes anurus* and *Microctonus aethiopoides* seems to have wrought the greatest success (Yeargan, 1985). The fungus *Erynia* sp. reduced alfalfa weevil numbers in several locations in Ontario (Harcourt *et al.* 1977; Nordin *et al.*, 1983). Harcourt *et al.* (1984) found that *Erynia* sp. and 18–57% sterilization due to *Microctonus aethiopoides* consistently kept the weevil below an EIL of 150 larvae/0.09 square meter. Several varieties of alfalfa resistant (via tolerance) to the weevil were developed and released into commercial production (Barnes *et al.*, 1970), where they joined over 30 varieties previously resistant to aphids.

The alfalfa ecosystem is subject to periodic disturbance by harvesting, and, in the northern states and Canada, management of alfalfa

TABLE 12-1

Parasitoids Imported into the United States for Biological Control of the Alfalfa Weevil[a]

Host stage	Species	First release	Range
Egg	*Patasson luna* (Mymaridae)	1911	General
Larva	*Bathyplectes curculionis* (Ichneumonidae)	1911	Most of United States and Canada
	Bathyplectes anurus (Ichneumonidae)	1962	Northeastern United States, southeastern Canada, Oklahoma, California[b]
	Bathyplectes stenostigma (Ichneumonidae)	1964	Northeastern United States (very local), Colorado
	Tetrastichus incertus (Eulophidae)	1960	Northeastern United States, southeastern Canada, California (rare)
Larva–adult	*Microctonus colesi* (Braconidae)	1966	Northeastern United States, southeastern Canada
	Microctonus stelleri (Braconidae)	1959	Very local, rare, perhaps not established
Pupa	*Dibrachoides dynastes* (Pteromalidae)	1911	Western United States (rare)
Adult	*Microctonus aethiopoides* (Braconidae)	1959	East and central United States, southeastern Canada, California[b]

[a]From APHIS–PPQ documentation.
[b]As of 1986; continued expansion of range is likely.

weevil by adjustment of harvesting date was often possible, destroying most larvae before significant damage occurred. Shoemaker and Onstad (1983) showed, via an optimization model, that early harvest was most effective and resulted in least disruption of biological control. In Wisconsin, McGuckin (1983) found that early cutting reduced alfalfa weevil numbers enough to increase returns by $61.75/hectare while reducing insecticide usage by 45%.

Ultimately, researchers developed management programs that embodied many of the principles of an ecological approach to pest management (Wilson & Armbrust, 1970; Armbrust & Gyrisco, 1982). Ruesink et al. (1980) developed a coupled insect–plant model in which plant matter (leaves and stems) was assumed to grow logistically with a developmental threshold of 5°C; a multicomponent alfalfa weevil population model including activity of the major larval parasitoid Bathyplectes curculionis was added on a developmental base of 9°C. Because plant and weevil populations were each temperature-dependent, a temperature-driven computer model gave accurate predictions of weevil population densities, especially when a preliminary count of early-instar larvae was input. Recently, growers have found it profitable to employ pest management scouts (usually supervised by the Cooperative Extension Service or by a private consulting firm) who sample weevil populations regularly in each field. In the Midwest, the critical data are: degree-day accumulation (base 9°C) since 1 January, alfalfa weevil larval density, and crop height.

Table 12-2 shows how EIL and treatment decisions vary as functions of these parameters. Censuses of weevil eggs and larvae are input into the computer model, which then produces an advisory for the grower. For instance, if alfalfa is 30 centimeters tall and 300 degree-days (base 9°C) have transpired since 1 January, and if a scout finds 105 or more late-instar larvae in a sample of 30 stems, the grower is advised to spray. If the scout finds 104 or fewer larvae he/she makes a return visit after 50 more degree-days. Insecticides are applied only when necessary, when the model predicts a high probability that the EIL will be exceeded. These insecticide treatments are timed for application when parasitoids are least likely to be active in the field.

Even as the alfalfa weevil management program has become a successful model of integrated crop management, the potato leafhopper (Empoasca fabae) has become a problem of increasing concern. The potato leafhopper sucks sap from a wide variety of agricultural and noncrop plants. In the northeastern and midwestern United States, the leafhopper becomes abundant in midsummer and causes yellowing, stunting, and loss of yield in the second and third alfalfa crop. Its damage is more insidious than that done by the weevil, for the

TABLE 12-2

Dynamic EIL for Alfalfa Weevil Management[a]

Degree-days (dd) since 1 January (9°C base)	Alfalfa height (average of 10 stems)											
	5 cm	7.5 cm	10 cm	13 cm	15 cm	18 cm	20 cm	23 cm	25 cm	28 cm	30 cm	33 cm[b]
190–210												
Apply spray	27[c]	47	67	85	100	115	130					
Sample again after 50 more dd	0-26	0-46	0-66	0-84	0-99	0-114	0-129					
240–260												
Apply spray	21	30	39	47	55	62	69	69	69			
Sample again after 50 more dd	0-20	0-29	0-38	0-46	0-54	0-61	0-68	0-68	0-68			
290–310												
Apply spray		25	37	52	67	75	83	94	105	105	105	
Sample again after 50 more dd		0-24	0-36	0-51	0-66	0-74	0-82	0-93	0-104	0-104	0-104	
340–360												
Apply spray					82	82	82	82	82	82	82	82
Sample again after 50 more dd					14-81	14-81	14-81	14-81	14-81	14-81	17-81	17-81
Sample again after 100 more dd					0-13	0-13	0-13	0-13	0-13	0-13	0-16	0-16
390–510												
Apply spray										52	52	58
Sample again after 50 more dd										8-51	8-51	8-57
Sample again after 100 more dd										0-7	0-7	0-7

[a]Modified from Wedberg et al. (1977).

[b]See Wedberg et al. (1977) for continuation of chart for alfalfa >33 cm, or sampling ≥540 dd, or after first harvest.

[c]Numbers are weevil larvae collected from 30-stem sample.

yield of protein is reduced significantly before yellowing is visible (Kouskolekas & Decker, 1968; Elhag, 1976). Biological controls are generally ineffective, and the search for resistant varieties has yielded equivocal results (e.g., Fiori & Dolan, 1981). Few farmers have the time or the expertise to sample every field regularly, so that IPM programs similar to those for the alfalfa weevil are under development for potato leafhopper (e.g., Flinn *et al.*, 1986). The EIL is a function of plant height (larger plants are more tolerant), and a temperature-dependent computer model has been developed (Onstad *et al.*, 1984). Despite the inaccuracies of the sweep net (Chapter 3), a trained pest management scout who samples regularly can estimate the relationship of the field population to the EIL and report on whether or not the threshold has been exceeded. For example, scouts in Virginia take three samples of 10 sweeps each, along with 10 random stems, and, using a sequential sampling plan (Table 12-3) have reduced sampling effort by 28–55%,

TABLE 12-3
Critical Values for Sequential Sampling of Potato Leafhopper in Virginia Alfalfa[a]

Plant height	Number of 10-sweep samples	Cumulative number of leafhoppers	
		No spray	Spray
0–18 cm	3	4	13
	4	7	16
	5	10	19
	6	13	22
18–26 cm	3	13	22
	4	19	29
	5	25	35
	6	31	41
26–35 cm	3	23	41
	4	34	51
	5	44	62
	6	55	72
35–43 cm	3	30	59
	4	45	74
	5	60	89
	6	75	104
>43 cm	3	35	69
	4	52	87
	5	70	104
	6	88	122

[a]Luna *et al.* (1983). © Entomological Society of America. Used with permission.

with an average error rate of only 1.8% (Luna *et al.*, 1983). Pest management scouting has led to greater awareness of the leafhopper and resulted in increased insecticide usage on alfalfa in the midwestern states, indicating that monitoring pest densities does not necessarily reduce the amount of pesticide applied to a crop.

Alfalfa tolerates wide temperature extremes, and can be grown in most seasonal climates where moisture, either natural or from irrigation, is adequate. In the arid southwestern United States, alfalfa grows throughout the year, and in southern California and southern Arizona it can be harvested up to nine or ten times annually. The pest complex in this region differs from that in the eastern United States. The key pest is the Egyptian alfalfa weevil, whose future densities can be predicted by using a temperature-dependent model very similar to that used for the alfalfa weevil in the eastern states. Optimal management of adults depends on estimating reimmigration of adults into fields after aestivation (Gutierrez *et al.*, 1976), and insecticide treatments are often necessary. The alfalfa caterpillar (*Colias eurytheme*) and several species of aphids are also capable of damage, whereas the potato leafhopper is rare and the impact of its feeding is negligible. Both alfalfa and its pests (weevil excepted) are present throughout the year, and strip-harvesting is a viable option. Segments of each field can be harvested alternately, so that there is always a standing crop to serve as a refuge for predators and parasitoids (Summers, 1976). Avoidance of broad-spectrum insecticides further conserves natural enemies, which are sufficiently numerous to control the alfalfa caterpillar and aphids most of the time. *Bacillus thuringiensis*, though expensive, may reduce local outbreaks of the caterpillar, again conserving natural enemies (University of California, 1981).

A unique pest complex occurs in seed alfalfa, where the major pests are *Lygus* bugs and the alfalfa seed chalcid. There is an overriding need to conserve pollinators, which are critical to the success of the crop, yet the major pests often are active during bloom. *Lygus* bugs can be controlled via careful application of short-residual insecticide to alfalfa late in the day, when bees are less active. Alternatively, some chemicals (such as metasystox-R), are repellent to bees (Haws & Davis, 1985). The seed chalcid is nearly impossible to control chemically, and more emphasis must be placed on cultural control. Growers within a region cooperate by cleaning up chaff and waste and feeding infested crops as hay to livestock before emergence of the next generation of chalcids. (The chalcid is often heavily parasitized by other Hymenoptera, though this does not help the seed producer, because the infested seed has already been destroyed.)

Alfalfa consistently attracts new pests, most recently (in North America) the alfalfa blotch leaf miner (*Agromyza frontella*) in 1972 and the blue alfalfa aphid (*Acyrthosiphon kondoi*) in 1974. The impact of these two pests is under investigation, although it appears that the aphid will present the greater problem despite the widespread use of resistant varieties of alfalfa. Gutierrez *et al.* (1984) have begun development of an energy flow model including alfalfa, pea, and blue alfalfa aphids, lady beetles, lacewings, and the aphid parasitoid *Aphidius smithii*. An initial stochastic simulation gave a good fit to field data, and showed not only that the natural enemies produce significant aphid mortality, but also that the phenology of the plant is important in enhancing predator activity. The alfalfa blotch leafminer is apparently under biological control via introduced parasitoids (Drea & Hendrickson, 1986).

VEGETABLE PESTS: INTENSIVE CROPPING WITH LOW EIL

An impressive variety of vegetables is grown for consumption by humans, and the pest complex associated with each is likewise diverse. As they do to most green plants, insects eat all parts of vegetables: leaves, stems, roots, fruits, and flowers, both in the field and in processing or storage. Damage done to vegetables by insect pests may be direct or indirect. Direct injury is to the portion intended for human consumption; for example, larvae of the tomato fruit worm hollow out ripening tomatoes, while those of the imported cabbageworm eat the wrapper leaves and then bore into the heads of cabbages. Indirect damage is done by insects' feeding on nonedible plant parts—for instance, leaves of potato by larvae and adults of the Colorado potato beetle, or leaves of beans by the Mexican bean beetle. Each pest reduces photosynthesis sufficiently to depress yields. Vegetable diseases can be transmitted by insects; notably, viruses and mycoplasmas can be vectored by aphids and leafhoppers, but also bacteria can be spread by chrysomelid beetles. A recent estimate is that vegetable growers lose $400 million annually to insects in the United States (Fronk, 1985).

Vegetable pest complexes exhibit many applications of the insect–plant associations detailed in Chapter 6. In most cases, there are both generalists and specialists among vegetable pests. General feeders include polyphagous species, such as cutworms or wireworms, or the green peach aphid, whose presence on over 300 host plants has been recorded (van Emden *et al.*, 1969), or the tarnished plant bug, which sucks sap from nearly all green plants. Specialists are closely associated with plant families; for instance, the Colorado potato beetle feeds on nearly all cultivated (and some wild) Solenaceae and is therefore likely to eat potato, eggplant, and tomato leaves with equal relish (though not with equal impact on the yield). Specialist members of the cabbage pest complex, such as the imported cabbageworm, diamondback moth, cabbage looper, and cabbage aphid feed on most species of Cruciferae.

Generally, each vegetable pest complex is dominated by a key pest, whose EIL usually is lowest because direct damage is done to the marketable portion of the plant and cosmetic injury is an important consideration. The key pest thus is dependent on the intended use of the crop. For example, both the imported cabbageworm and the cabbage maggot eat Cruciferae, feeding, respectively, on leaves and roots. The cabbageworm is the key pest of cabbages because of cosmetic damage to leaves, whereas the maggot is the key pest of rutabagas due to cosmetic damage to roots.

Management of insect pests and diseases of vegetables is undertaken in the context of very low EILs for the commercial grower. In the United States and most other economically developed nations, consumers demand blemish-free produce, and the extremely low EILs faced by commercial growers of fresh produce present a problem in management of vegetable insect pests. Pleasing appearance is considered synonymous with "quality." Consumers' desires are reflected in regulations governing maximum allowable amounts of insects and their fragments in both fresh and processed foods (Pimentel et al., 1977). Although EILs are slightly higher for processed than for fresh vegetables (e.g., cabbage for sauerkraut), increased use of processed vegetables in the convenience-food industry has led recently to a steady reduction in EILs for vegetable pests (Bundy, 1977). Commercial producers of vegetables, in order to return a reasonable profit on their investments, are faced with the necessity of controlling pests at levels far below those that might result from natural or biological control. Routine applications of broad-spectrum insecticides are seen as essential under most current marketing conditions, in order to protect investment. This is expensive and leads to insecticide resistance, and some vegetables (notably cucurbits) are susceptible to insecticide-induced phytotoxicity. (WPs are sometimes preferable to ECs in vegetable insect control because of the tendency of ECs to "burn" plants at high ambient temperatures.)

Indiscriminate insecticide application and its undesirable side effects (Chapter 7) add to costs of control. The minimum task of the applied entomologist is thus to improve efficiency of insecticide application by modifying spray equipment to avoid waste, and by developing more precise timing of insecticide applications. ULV formulations and highly accurate nozzles provide more even coverage, and a greater proportion of insecticide is likely to contact the target insects. Spraying according to calendar date, now often done whether or not pests are present, can be adjusted by efficient monitoring of pest populations. Periodic monitoring by inspection of plants or by pheromone traps can warn vegetable growers of insect infestations just as in alfalfa pest management programs, and some vegetable industries have initiated similar pest management scouting programs. For example, one IPM program for cabbages makes use of "cabbage looper equivalents," 1 cabbage looper being the equivalent of 1.5 imported cabbageworm or 20 diamondback moth larvae. (This reflects the relative sizes, and consequent damage wrought by, these three species.) Andaloro *et al.* (1983) reported that a 50% reduction in insecticide usage resulted from sampling cabbage looper equivalents on 40–50 plants in a V-shaped tour through the field. In another case, Wyman *et al.* (1977) found that emergence of cabbage maggot (*Hylemyia brassicae*) adults was closely correlated with cumulative temperatures. Insecticide applications timed by phenological date rather than by calendar date were more effective in reducing infestations, and less insecticide was used.

EILs for vegetable pests are variable, depending upon circumstances. For green peach aphid on seed potatoes, Cancelado and Radcliffe (1979) used an EIL of 10 apterae/105 leaves, which value rose to 30/105 for market potatoes. The economic threshold (and therefore the EIL) for lepidopteran eggs on cabbage is higher if a chemical insecticide is to be used than if BT is to be applied (Sears *et al.*, 1983). Both are equally effective, but BT takes longer to elicit an impact. Employment of computer-simulated population models in vegetable IPM is under development. One of the first such models, "GPA–CAST," uses results from preliminary samples of green peach aphid (starting from a base of 5 aphids/50 leaves), predicted temperatures, and market conditions. These are used to generate advisories, for growers, that reveal the dollar-value potential to be realized from each management option (Smilowitz, 1981). Such management programs will probably be routine procedures for computer-literate farmers within the next few years.

If pests and their damage are to be monitored, it is important that growers or scouts be able to recognize specific pests and their damage. For instance, when populations of the spruce budworm are high in the forests of Maine, the moths may appear in large numbers in surrounding potato fields. They do not damage potatoes, but some growers spray regularly and repeatedly when confronted by clouds of moths in their crop. These treatments are needless and wasteful of insecticide. Similarly, the Colorado potato beetle feeds on leaves of tomatoes but causes very little yield loss (Cooper, 1981). Tomato growers in the American Midwest must be aware of this and must refrain from spraying insecticides unnecessarily.

Noninsecticidal management techniques for vegetable pests are varied, although low EILs have impeded their development. BT is available for management of larval Lepidoptera and is used extensively on cole crops; its advantages are that it can be applied very close to harvest and that it does not disrupt biological control of other species (Sears *et al.*, 1983). The expense of BT is not as great a contraint in crops of high value/hectare, such as vegetables, as it is in systems of lower relative cash value. Cleanup of crop residues and postharvest plowing greatly reduce overwintering populations of such pests as squash vine borer, tomato hornworm, and Colorado potato beetle. Planting arrangements may have a decided impact on pests; planting beans 7–10 centimeters apart, rather than the customary five or fewer, reduces Mexican bean beetle impact, with no concomitant yield loss (Fronk, 1985). Many insect-resistant varieties of vegetables exist (G. Kennedy, 1978), and development of this potential is continuing. There is partial biocontrol from many introduced natural enemies, and this might be enhanced through suitable manipulations. For instance, Gould and Jeanne (1984) found reduced populations of cabbageworm and looper when density of *Polistes* wasps was increased by provision of nesting sites. Vegetable pest management systems, with their inevitable use of insecticides, are an ideal candidate for the release of genetically improved predators that resist insecticides (Chapter 9).

EILs used by home and community gardeners are variable and are usually higher than those determined by commercial growers. Cosmetic standards of home gardeners run the gamut from absolute tolerance of any and all insects to tolerance of none. In labor-intensive operations, such as home gardens, it is possible to hand-pick insects or to cull or trim (or simply to eat) blemished produce. The home gardener does not have to show a profit (in the garden) to stay in business, and if a crop is severely damaged or destroyed, the gardener can usually go to the grocery store for vegetables. Also, the home gardener can adjust for potential insect damage by planting slightly more than anticipated needs, assuming space and labor are available. (In my own garden, a 20% increase in area planted has offset nearly all losses to insects and disease, and the investment has been less than that for using insecticides, though I have not factored cost of labor into my analysis.) Routine monitoring to spot incipient infestations, and a close watch thereafter to see whether yields are likely to be reduced, aid in limiting investment in pest control techniques.

Home gardens typically are small and diversified plots, in contrast to the monocultures demanded by mechanized commercial production. Intercropping (Chapters 6 and 11) can be used effectively in a home garden, and increases plant diversity along with habitat heterogeneity, to reduce the potential increase of phytophagous specialists. The surrounding habitat may provide shelter for potential agents of biological control, which can be enhanced by limiting chemical applications to spot treatments of infested plants only.

COTTON PESTS: DEVELOPED VERSUS DEVELOPING ECONOMY

Forty percent of insecticide applied in the United States is used on a single crop, cotton (Eichers, 1981; Pimentel & Levitan, 1986). Cotton occupies up to 8 million hectares in the southeastern and southwestern United States and 25 million additional hectares in 80 nations worldwide. The United States, India, China, and the USSR are the world's leading producers. Cotton varieties under production are derived from four ancestral species—two from tropical America and two from the Middle East—an exception to the general trend of a single presumed origin for major crop plants. Cotton is grown for both fiber and oilseed, and an advantage is that the crop has a low water demand, which makes it suitable for semiarid climates. Annual insecticide applications to cotton in the United States cost from $200 up to $700/hectare (Reynolds et al., 1982), despite which an estimated 10% of the crop is lost annually to insect attack. This heavy and repeated insecticide use has been deemed necessary to reduce densities of a large pest complex of about 125 damaging species, dominated by a few key species.

The key pest in much of the United States is the boll weevil (*Anthonomus grandis*), which made its first appearance in Texas during the 1890s and by the late 1920s had spread throughout much of the cotton-producing region of the southeastern United States. (It is still spreading, though very slowly, in the High Plains region of Texas.) Boll weevil adults overwinter in sheltered sites, usually near cotton fields. Emerging adults move to cotton, where they feed by puncturing the developing floral buds ("squares"). Eggs are deposited in the squares, within which the larvae feed and cause premature abscission. Later, both adults and larvae feed upon the ripening fruits ("bolls"), and destroy the seeds and developing lint before harvest.

There are two to seven annual generations. Insecticides directed against the boll weevil, especially following the widespread adoption of synthetic organic chemicals, have resulted in outbreaks of secondary pests, notably *Heliothis* spp., the cotton bollworm, and tobacco budworm, each of which feeds on a wide variety of crop plants, including corn, soybeans, tobacco, and tomato. Additionally, the pink bollworm (*Pectinophora gossypiella*), originally from Asia, is a pest in the semiarid irrigated cotton-producing areas of the southwestern United States. Spider mites have become a major secondary pest after intensive use of insecticides. About 25 species of cotton pests have developed resistance to insecticides. The boll weevil was among the first (by 1955); the pink bollworm, *Heliothis* spp., and several species of aphids all resist insecticides in portions of their geographic ranges.

At the same time that heavy insecticide use brought on resistance and the outbreak of secondary pests, it was recognized that insecticide usage could be significantly reduced through use of integrated management techniques. Cotton was the first crop routinely scouted for insect damage (in Arkansas, commencing in 1901). In some instances, pest management scouting has resulted in reduction of sprays on cotton from as many as 18 annual applications to as few as two, with no apparent loss of yield. Smith *et al.* (1974) found that Mississippi cotton growers who monitored pest damage spent an average of $39.93/hectare, a savings of $35.56/hectare over growers who used a preventive program of weekly sprays.

Adoption of scouting in Texas reduced insecticide input from 6.14 kilograms/hectare in 1976 to 1.68 kilograms/hectare in 1982 (Frisbie & Adkisson, 1985). Pest management scouts in Texas sample once or twice weekly and recommend spraying when 15–25% of squares or bolls show feeding punctures or when there are 40 fleahoppers per 100 terminals. Arkansas scouts use an economic threshold of 1.0–1.5 punctures squares per foot-row (Cate, 1985). In California, pink bollworm densities are monitored via pheromone traps baited with hexalure; 3.5–4.0 bollworms/trap constitute a population capable of damage, and the grower is advised to initiate spraying (Pfadt, 1985). In general, cotton growers have been slow to adopt crop scouting (Reynolds *et al.*, 1982).

Population models for the cotton ecosystem have used coupled plant–insect submodels. Plants have been modeled using leaf, stem, fruit, and photosynthate as separate components, as all these are affected by insect attack. Modeling the cotton plant is a challenge because of its unpredictability; growth of cotton is indeterminate (Gutierrez *et al.*, 1980). A temperature-dependent model developed in Texas considers a stochastic function about an "average plant," and shows that small changes in the plant are enough to reduce boll weevil densities to below the EIL (Curry et al., 1980). Models for cotton pests in California are multicomponent life tables in which age structures are used to simulate population changes (Chapter 4; Gutierrez et al., 1977b; Wang et al., 1977). In these models, pink bollworm population performance is a function of food quality. Models currently exist for the boll weevil, for several Lepidoptera, including the pink bollworm and Heliothis, and for spider mites and Lygus bugs. "BUGNET" is a pest management program currently on-line in Texas and has contributed to significant reduction in insecticide usage there. COMAX is a cotton management system including a weather-driven plant simulation model that relies on artificial intelligence to provide recommendations on irrigation, fertilization, pesticides, and harvest (Lemmon, 1986). Using COMAX, growers saved up to $60/acre in production costs. In Australia, sequential sampling of Heliothis spp. and spider mites thrice weekly, coupled with a computer model, has resulted in a 40% reduction in insecticide usage on cotton (Hearn et al., 1981).

There is considerable economic pressure to develop pest management programs for cotton, because the crop competes economically with synthetic fibers that can often be manufactured with less expense than that of growing and processing cotton. In the United States, cotton producers are assessed $1/bale, which contributes to a fund promoting both the use of cotton and alternative strategies for insect management that might reduce heavy dependence on insecticides. In a wealthy nation such as the United States, economic resources are available to subsidize the development of alternative management strategies to the benefit of producers and society at large. For instance, NPV (Chapter 8) is available for Heliothis management, due in part to government subsidy of its development (Chapter 8). NPV is most effective when applications are timed to coincide with Heliothis egg hatch.

Resistant varieties are under development (Niles, 1980). An example is frego-bract cotton (Chapter 10), in which a single gene results in antixenosis to boll weevil, though susceptibility to Lygus spp. increases slightly. Pubescent and red varieties also resist boll weevils somewhat (Gallun & Khush, 1980). Varieties high in gossypiol concentration resist both Lygus and Heliothis spp.

Other noninsecticidal management techniques for boll weevil include uniform planting dates after peak weevil emergence in the spring. Growing and prompt harvesting of early-maturing varieties greatly reduce infestations of boll weevil and pink bollworm (Masud et al., 1981). Frisbie and Adkisson (1985) reported that planting short-season varieties avoided insecticide costs of up to $117/hectare. Some additional control may be had by clean tillage of crop residue, which destroys weevil-infested bolls (along with other insects) late in the season and may reduce the weevil population entering overwintering. An eradication program is underway against boll weevil in the southeastern United States. Insecticide is used on a regional basis to reduce populations to levels at which releases of sterile males might be effective. Pheromone traps are used to lure male weevils to a bait containing a potent chemosterilant. After being thus rendered sterile, the weevils mate, and females lay inviable eggs. Preliminary results have been much more favorable on the periphery of the weevil's range than in the center: reimmigration rate is much less in North Carolina (on the fringe of the infestation) than in Mississippi.

Several species of predators and a few parasitoids consume quantities of cotton pests, though their numbers are often too low to control the pests adequately (Clausen, 1977). Three parasitoid species (*Peristenus* spp.) from Europe have been released against the tarnished plant bug in Mississippi (Phillips *et al.*, 1980), though their impact has been minimal. Cate (1985) reported that the imported fire ant may be a factor in keeping weevil numbers low—a mixed blessing, indeed. The recent reductions in insecticide use on cotton enhance the prospects for effective biological control (Pfadt, 1985).

Regional management with cooperation among growers shows promise in management of bollworms on cotton by destruction of wild host plants (Knipling & Stadelbacher, 1983). Community-wide timed sprays help against pink bollworm (Phillips *et al.*, 1980).

Cotton pest management in a developing nation lacking sufficient resources for a broadly based management approach presents a contrast with the situation outlined above. For example, in the Democratic Republic of the Sudan, nearly a million hectares of cotton are grown, mostly under irrigation, and cotton is the chief export crop. A large share of the Sudan's foreign exchange is generated from this single crop, and failure in insect management, and consequently of the crop, would spell a major economic disaster that might threaten the survival of the Sudan as a nation. Economic resources within the country are far more limited than are those in the United States. There is a large and diverse pest complex (Ripper & George, 1965), and cotton production in the Sudan depends on regular and repeated heavy treatment with persistent broad-spectrum insecticides applied up to 30 times annually. The choice of chemical is most often based on minimal expense. The least expensive insecticides are chlorinated hydrocarbons (including the extremely persistent cyclodienes) and the most hazardous of the organophosphates, which may be applied by untrained personnel, to whom accidents may occur. Moreover, large quantities of insecticide residues flow down the Nile River through Egypt and into the Mediterranean Sea, potentially contaminating fisheries there. Some spectacular cases of resistance and secondary pests have developed. For example, whiteflies, normally not a major problem on cotton (or on most other crops outdoors), became a secondary pest after their predators and parasitoids were eliminated by insecticides. Whiteflies suck sap and cause wilting. They also produce honeydew, which evaporates very quickly under the Saharan sun, leaving a residue of sugar crystals. These crystals remain as an impurity in the cotton lint, and at high concentrations they cut strands of lint and render the cotton useless for textiles.

Alternative pest management strategies are needed if cotton production in the Sudan is to avoid the crisis and disaster phases described in Chapter 1. Kumar (1984) reported some progress in identifying EILs for whiteflies and jassids (Cicadellidae, mostly *Empoasca* spp.). All cotton residue after harvest must be destroyed (by law) for management of pink bollworm. Early planting also aids in avoiding heavy infestations of bollworms (Ripper & George, 1965). Management of weeds near cotton fields reduces populations of pests by eliminating wild hosts. Resistant varieties would give the most effective lasting control, ideally with a mixture of varieties to take advantage of the wide climatic variation and to forestall the evolution of biotypes (Joyce, 1959; Mound, 1965).

Without the resource base to support a diversified research establishment, and with but a single major cash crop, the Sudan is hardpressed to develop solutions to the serious insect pest problems involved in cotton management, despite an apparently shortsighted dependence on broad-spectrum insecticides. Additionally, the Sudan shares its borders with eight different and diverse nations, not all of whose governments have been paragons of political stability or friendliness. Some of the limited foreign exchange that, under more tranquil political conditions, might pay for agricultural research and development is diverted to maintaining a military establishment of sufficient strength to deal with potentially hostile neighbors.

FOREST INSECT PEST MANAGEMENT: HIGH VARIABILITY AND DIFFICULTY IN ASSESSMENT OF EIL

Improvement of insect pest management in forest ecosystems presents some unique challenges, whether the forest is "natural" or intensively managed. Most forest ecosystems, except for carefully and intensively managed single-species tree farms, contain a high diversity of plant species, microhabitats, and insects, and targeting management against any single pest species becomes difficult because of this complexity (Coulson & Witter, 1984). As an example of forest insect diversity, I have collected over 130 species of noctuid moths in a single blacklight trap in a temperate deciduous forest near Columbus, Ohio, on alternate weekends since 1975. A malaise trap at the same location has yielded over 350 species of Ichneumonidae. Even this diversity is modest when compared to the bewildering variety of insects within tropical rain forests, a single hectare of which may harbor over 10,000 insect species (Erwin, 1983). Forest ecosystems generally display temporal stability, in comparison with most agricultural crops. The frequent total disruption typical of planting and harvest operations in agricultural ecosystems is generally lacking in most forests, where harvest or disaster generally occurs at rather infrequent intervals.

All parts of trees are attacked by a great variety of insect pests (Johnson & Lyon, 1976). In coniferous forests, the most important pests (usually) are several species of Scolytidae (bark beetles), notably *Dendroctonus* spp. and *Ips* spp. Defoliators such as the spruce budworm, Douglas-fir tussock moth, and sawflies are also major pests because of the particular sensitivity of conifers to defoliation (Chapter 6). Lepidopteran defoliators such as the gypsy moth are among the most severe pests in deciduous forests (Thompson, 1985), though woodborers (Cerambycidae, Buprestidae, and Sesiidae) can also devastate deciduous trees. All these pests often kill trees outright, and many other pests of lesser impact occur. These weaken trees, slow their growth, and increase susceptibility to disease, fire, wind, or ice.

Coulson and Witter (1984) characterized three general kinds of forests: (1) multiple-use "natural" forests, (2) specialized intensive tree farms, and (3) so-called "urban forests." Approaches to pest management differ somewhat within these three classes. In forests managed for multiple uses, wood and other forest products are harvested under management regimes of varying intensity. Timber management, grazing, recreation, wildlife conservation, and watershed protection are land uses established by statute in the National Forests of the United States, though there is a tendency to emphasize one or two of these uses in local management. A "natural" forest may also be permanently inhabited by homeowners, whose numbers are increasing (in the United States) due to emigration from urban and suburban areas. Truly "natural" forests, unmanaged by humans, are very rare indeed, and, by definition, such wilderness areas do not require that insect pests be managed.

The principles of tree farming are closer to those of agricultural crop management than to those of traditional forestry. Tree farms are specialized, often single-species, even-aged plantings, and are an exception to the general rule that forest ecosystems display high species diversity. In tree farming, unwanted plant species are considered pests, and are controlled, often via herbicides. Frequent use of heavy machinery compacts the soil and may injure trees. Usually, a monoculture results in a smaller pest complex, often displaying higher population densities (Chapters 6 and 11). The higher value of trees on tree farms results in a lower EIL for key pests. Decision-making is less complex than in multi-use forests because only a single landowner or manager may be involved. Tree farms are usually privately owned (in North America, western Europe, etc.), and the areas needing management are often smaller than extensive public forest lands, such as the U.S. National Forests.

In the "urban forest," trees are planted in parks, along streets, and for landscaping in both public places and private areas. These trees are aesthetic and also ameliorate the impact of weather—particularly, excessive sun and wind. A higher proportion of exotic species is present than in corresponding "natural" forests; often, species diversity is high and many age classes are represented. Monoculture in the urban forest has led to disaster on occasion. The outstanding example (in North America) has been the near-elimination of the stately American elm as a shade tree due to the depredations of Dutch elm disease, transmitted by elm bark beetles. Stresses are often increased. Unique hazards range from industrial and automotive air pollution to wayward vehicles and vandalism. Soil disturbance and compaction due to construction and maintenance of utility lines, roadways, and so forth are additional sources of stress. Drought, especially in semiarid and arid climates, adds significantly to insect and disease susceptibility. Simultaneously, close proximity of trees to people in the urban environment causes greater emphasis on aesthetic injury levels, the so-called "dead leaf syndrome" (Coulson & Witter, 1984). There are homeowners who tolerate not so much as a single insect's biting of a leaf, despite evidence that minor phytophagy is routine and does no lasting injury to the plant (Chapter 6). Finally, venerable historic (and therefore senescent) trees are often particular favorites, and local officials may go to great lengths to preserve them from natural and normal insect attack. For many years, the stump of the "Washington Elm" stood in commemoration of General George Washington's 1776 visit to the Common at Cambridge, Massachusetts, despite the tree's being riddled with borer holes. Aged sycamores in Columbus, Ohio, are sometimes afforded similar respect, despite the occasional hazard of huge limbs' dropping onto parked automobiles.

The diverse nature of forest ecosystems, plus their multiple uses (tree farms excepted), causes conflict in decision-making for pest managers. The gypsy moth (*Lymantria dispar*) is an example. It is a cyclic defoliator of hardwood trees (oaks are preferred), and was introduced from Europe into the northeastern United States, from whence it has slowly spread southward and westward. Gypsy moths have a single generation annually and populations irrupt at intervals of about 7 to 10 years. An outbreak may last 2 or 3 years, causing total defoliation during late June and early July in infested regions. There is limited mortality among white oaks, and most other tree species are stressed but survive. Low-intensity mortality may thin a forest to the extent that timber production is actually improved (Stark *et al.*, 1980). Such cyclic irruptions of forest defoliators are generally common among both exotic and native species, and are a "natural" factor in forest dynamics (Anderson & May, 1980). (The gypsy moth is not "natural" in North America but it undergoes cyclic irruptions in its native Eurasia as well. Most forest insect pests in North America are native, and many of the insect pests and pathogens of urban forests are of the same species as those of "natural" forests.)

Difficulties arise in determining EILs for gypsy moth, and these are compounded when continuous forest owned by a patchwork of public and private interests is managed for varying uses. The home-owner within a gypsy-moth-infested woodland is very likely to display especially low tolerance for the insects. The specter of defoliation is bad enough, but the sight of the large and hirsute caterpillars and the nightly rain of frass are of even greater concern to the resident land-owners. Moreover, allergy to larval setae is quite common and causes much unpleasantness as exuviae accumulate during an outbreak. A similar level of intolerance occurs in recreational forests, especially around campgrounds. Despite a yearning to be in the "wilds," many American campers desire "wilderness" on their own terms, preferably insect-free. Economic loss as a result of mountain pine beetle in Idaho was estimated at $2 million when campground occupancy for heavily infested and uninfested regions was compared (Michalson, 1975).

On the other hand, commercial foresters primarily interested in management for timber production might wish to delay any overt, expensive management activity until the outbreak reaches proportions that actually interfere with tree health. If oaks are ready for harvest anyway, a prudent timber producer might welcome the depredations of the gypsy moth so as to reduce the amount of foliage with which he or she must deal after harvest. If the forest were part of a natural preserve, or a designated wilderness, there might be agitation for doing nothing except letting nature take its due course. Differences of opinion become acute when both public and private ownership and management are involved, as when a wildlife sanctuary abuts a wooded tract studded with $300,000 homes. About 70 species of parasitoids and predators have been introduced in an effort to control the gypsy moth biologically, and while many have become established, a lasting reduction in gypsy-moth numbers has yet to be documented (Doane & McManus, 1981).

Management tactics against forest insect pests need to reflect the vast complexity involved in forest pest management. There must be rational determination of EILs and predictive models to forecast the impact and extent of outbreaks (Berryman & Stark, 1985). Many population models have been developed for major forest pests, and adequate simulations exist for spruce budworm, gypsy moth, Douglas-fir tussock moth, and several species of bark beetles on conifers. The most reliable models incorporate submodels for pest, treatment, and impact into an overall model of forest dynamics (Thompson, 1985). For example, a predictive model for southern pine beetle is adequate in forecasting, despite a high variance (Coulson *et al.*, 1980), and it is used extensively in tree farm management.

Forest insect problems must be considered over an extremely long time frame, with the unpredictability that implies. For example, the red oak borer infests young trees 20–30 years old, and economic damage is apparent only when the wood is harvested, which may occur up to a half-century later (Chapter 2). Is a woodlot owner better off investing his money in mutual funds rather than using it to assure timber of higher value in 50 years, assuming lack of major windstorms or other tree-destroying environmental disruptions?

Even if the economic impact of insects is adequately assessed, it is sometimes difficult to determine accurately the extent of damage or the size of insect populations in remote areas. Habitat heterogeneity makes adequate estimates of density hard to come by. Routine forest insect surveys are helpful in this assessment, and remote sensing via aircraft or satellite is very effective in evaluating insect infestations over remote and relatively inaccessible regions. Success in managing outbreaks of forest pests is greatly dependent on early detection.

When at all possible, broad-spectrum insecticides should be avoided in managing forest ecosystems. Aerial application is the only practical technique for chemically treating wide areas, and the hazards of drift are well-known (Chapter 7). By far the best strategy for forest pest management is long-term prevention of infestations. Führer (1985) and others have pointed out that endemic long-term regulation of forest pests is primarily due to large complexes of natural enemies, coupled with maintenance of vigorous, healthy stands. Selective cutting and pruning maintains trees in good health and can forestall outbreaks of many of the most devastating forest insect pests, especially in the "urban forest." In commercial forests, the chance of reinfestation of logged-over areas can be reduced by management of logging debris, including as little disturbance as possible, and removal of slash and debris. Reforestation with a diverse mix of trees is also helpful, and fertilization may be necessary initially to replace nutrients lost when slash is removed. On an experimental basis, it has proven feasible to "trap out" the western pine beetle, at least locally (Wood, 1980).

Though genetic heterogeneity is high among outbreeding forest trees, the long-term nature of forest systems makes the use of resistant plant varieties an attractive possibility (Hanover, 1980). Examples include development of urban elm trees that are partially resistant to Dutch elm disease and elm bark beetles. Moderate resistance to locust borer has been developed in black locust, and some commercial varieties of Norway spruce resist the Eastern spruce gall aphid. European scientists have developed late-flushing varieties of spruce that avoid heavy infestation by tussock moths and sawflies.

SUBURBAN HOUSEHOLDS: SUBJECTIVITY AND AESTHETIC INJURY LEVELS

Throughout this book, there has been mention of subjectively determined aesthetic injury levels, in which psychological factors interplay with economics and health to produce a great range over which people perceive insect pest problems. Urban pest management is a burgeoning specialty involving public education, estimation of aesthetic injury levels, and studies of insect ecology in the urban environment. Recent useful compendia of the state of the field are those by Frankie and Koehler (1983) and Bennett and Owens (1986). In general, arthropods are not tolerated well by 20th-century Americans, and probably most urban and suburban persons are unwilling to accept the low densities of insects that may result from ecological approaches to pest management (Byrne et al., 1984). As a consequence, insecticides are probably more overused and misused in domestic urban and suburban settings than in agricultural, silvicultural, or public health operations. For instance Bennett et al. (1983) found that up to 50% of homeowners (in Indiana) using a pesticide did not read the label. Pimentel and

Levitan (1986) estimated that half of all insecticide purchased by homeowners was disposed of in municipal trash.

Limited sampling reveals that urban residents display an appalling lack of basic knowledge of insect management. Wood *et al.* (1981) found that only 14% of residents in public housing (Maryland and Virginia) realized any connection between sanitation and cockroach infestation. Sawyer and Casagrande (1983) noted that attitudes toward arthropods were largely conditioned by cultural values. This suggests that successful urban insect pest management may depend as much on educating the public about the particulars of arthropod biology and safe pesticide use as on development of ecosystem models and integrated pest management systems, such as those in use by commercial agricultural operations.

Nowhere is this more evident than in the "average" household in the United States. An "average" home or homeowner probably does not exist outside the minds of pollsters. The example detailed in this chapter is simply one chosen for convenience and out of experience, though it is typical of single-family dwellings in a city considered by demographers to represent a median cross section of American humanity. Columbus, Ohio, is a major test market for new products; what is accepted in Columbus usually sells throughout the United States and, eventually, most of the world (not always to the total benefit of humanity).

In the present example, a three-bedroom two-story frame house occupies a landscaped lot 15 × 45 meters. What is not occupied by the dwelling, garage, or asphalt driveway is mostly a lawn of mixed grasses interspersed with a few short broadleaf weeds, and bordered by flower beds and ornamental shrubbery. There is a small vegetable garden. The only large woody species, candidates for the "urban forest," are two senescent blue spruces, a red maple, and a pear tree. The overall arrangement is typical of older (50+ years) dwellings in Columbus and elsewhere in the midwestern United States. What is atypical is a greater-than-usual tolerance for insects on the part of the inhabitants, though not all members of the author's family, nor his guests, share his enthusiasm for living arthropods.

Since 1973, over 800 insect species have been collected in house and yard, and the survey is by no means exhaustive. An entomophobe might consider all 800 or more to be pests, though only 33 are considered pestiferous enough to have been given English names by the Entomological Society of America. Of these 33 recognizable "pests," 16 have remained consistently below the author's EIL, and no action is taken against them. These "nonpests" include the following.

- Cucumber and asparagus beetles, Mexican bean beetle, wire-worms, whiteflies, *Lygus* bugs, and the imported cabbageworm all occur on or in vegetables, yet yields of edible fruits and roots have consistently exceeded the appetites of the household.
- Grape berry moths and codling moths occur on or in grapes or pears, but less than 10% of fruit is damaged. (Squirrels eat most of the pears before ripening, and the grapes, moth larvae included, are transformed into wine.)
- Houseflies and stable flies are kept at bay via screens and judicious use of a fly swatter.
- Cockroaches (German and Oriental) occasionally are introduced via cartons or cellar flooding. They are mashed as encountered and have failed to establish a thriving population. General (though not compulsive) tidiness in food-handling is a factor in preventing successful colonization.
- Paper wasps (*Polistes* spp.) nest under eaves, though not in densities high enough to be bothersome. They are evicted before painting or repairs. (Colonies near the vegetable garden seem to aid in keeping cabbageworms at low density, though a controlled experiment has not been done.)
- Carpenter ants infest fenceposts (owned by a neighbor) but not the house or garage, whose wood is too dry, owing to sound roofing and general good repair.
- In the lawn, sod webworms occur in low numbers; they are heavily parasitized.

Most pests of this particular household are dealt with individually rather than in an ecologically based integrated pest management system, yet integrated management of the complex is implied by some of the considerations mentioned: screening, general tidiness (especially in handling food and garbage), and early detection and elimination of invaders (such as cockroaches). Specific management tactics have been and continue to be used against the remaining 17 pests, as follows.

Pests Affecting Human Health

Mosquitoes of three genera (*Aedes*, *Anopheles*, and *Culex*) are common from March until October. Their bites are bothersome, and, in central Ohio, *Aedes* and *Culex* spp. are vectors of viral encephalitis. It is necessary to undertake an active and vigorous program of mosquito management, besides the aforementioned window screens and payment of taxes to finance county mosquito abatement programs. Standing water (in sagging eave spouts or sandbox toys) is drained when possible, and tall brush (wherein adult mosquitoes rest) is cut away from near the places in which adult humans rest. Repellent is used liberally during midsummer. An informal educational program for nearby neighbors probably helps; all the neighbors know that mosquito larvae develop in standing water and that adults rest in tall vegetation.

Yellow jackets (*Vespula* spp., primarily *V. germanica*) occasionally nest in the walls of the house. Their foraging probably aids in reducing caterpillar densities in yard and garden, though they are bothersome and sting occasionally. Because the owner's daughter is sensitive to the venom, the wasps are normally eliminated via insecticide or harvesting.

The human head louse has returned from school on occasion and requires a timely application of the appropriate insecticidal shampoo. This involves minor expense and inconvenience, and generates minor embarrassment.

Structural Pests

The eastern subterranean termite is common in the neighborhood, and, if unchecked, might do serious structural damage to house and garage. Professional treatment of foundations with heptachlor (last undertaken in 1968) has prevented recent infestation. Scrap lumber and firewood are kept off the ground and away from the house.

Closet and Pantry Pests

Black and brown carpet beetles are present chronically in low numbers within crannies containing 55 years' accumulation of wool dust from carpeting and furniture. Fumigation would eliminate them but is considered too expensive, so they are managed by dry cleaning of woolens before storage in airtight (and therefore beetleproof) plastic bags with paradichlorobenzene as a space fumigant. Insecticides are used to spot-treat potential sources of infestation (baseboards around the floor, and the interior of the piano, where carpet beetle larvae feast upon the felt pads).

Indian meal moths appear as if by magic in dried foods. Their management is simple: discard the infested item. Locating the primary focus of infestation sometimes presents a challenge worthy of a Sherlock Holmes or Hercule Poirot. Pet food, unemptied vacuum-cleaner bags, and artistic collages of beans or corn for school projects are prime foci, along with private hoards of stale candy from parties of the past decade.

Vegetable and Fruit Pests

Obviously, the pest complex on vegetables varies depending on the specific plants in the garden (usually squash, tomato, radishes, bean, broccoli, beet, carrot, pea, lettuce, asparagus). As mentioned earlier, several species considered "pests" in commercial production occur at densities lower than are justified by the homeowner's taking control measures. The following species do sometimes stunt or kill plants, reducing yield to the point that chemical control is deemed justifiable: flea beetles (radish and seedling broccoli), cutworms (seedlings of all kinds), tomato fruit worm, and squash vine borer. All but the last named have been readily controlled via application of the appropriate insecticide on an as-needed basis (EIL corresponds to the wilting of the first plant). Insecticides have been ineffective against the squash vine borer, despite experimentation with timing and dosage. This is a mystery, though resistance is a possibility. Recently, squash has been planted only in alternate years to avoid buildup of the vine borer.

Ornamental and Lawn Pests

As noted earlier, the owner has an unusually high tolerance for arthropods, reflected in a high aesthetic injury level. Insecticide is used sparingly on ornamental plants and only when they are visibly stressed to the point of wilting or yellowing. The only pests of ornamentals that have provoked action have been cottony maple scale (dormant oil was applied twice), spruce spider mite (dicofol was applied once), and Japanese beetle (carbaryl dust was applied once to roses). Insecticides have never been used on the lawn. Neighbors on all sides apply insecticides regularly and repeatedly. None of the neighbors has ever seen a chinch bug or a sod webworm. All lawns remain uniformly green, whether sprayed or unsprayed, except during prolonged drought, when they all turn brown together.

(Two vertebrate species are considered pests. The house sparrow is a potential reservoir of viral encephalitis, and its droppings are unsightly, especially on red shingle siding. Nests behind loose boards are eliminated and cracks are sealed as discovered. White-footed mice sometimes move indoors during November and they are eliminated via traditional mousetraps. A lean and hungry cat could easily control both biologically, but the cat in residence has not seen fit to do so.)

CONCLUSIONS AND FUTURE PROSPECTS

A common theme stated or implied throughout this book and developed in the five examples in this chapter is that an accurate assessment of pest impact is essential to implementation of ecological approaches to pest management. In agroecosystems especially, there is a need to translate estimation of pest impact into usable EILs. An ecological approach to pest management requires that disruptive actions be taken only when no other option is available. Accurate definition of economic and aesthetic injury levels is probably the major challenge to successfully combining simulation models with recommendation schemes based on expert systems and artificial intelligence.

Ecological approaches to pest management in ecosystems other than the five preceding examples bear many similarities, although every agricultural, urban, and public health pest management system has unique aspects also. In nearly all instances, however, insect pest management should be considered a part of total environmental management if long-term control is to be successful (Risser, 1985). The days of addressing insect problems separately from other aspects of ecosystem management are numbered, along with heavy-handed reliance on broad-spectrum insecticides. An ecological approach to pest management implies additional development of narrow-spectrum management techniques such as those discussed in Chapters 8 through 11. Novel techniques are desirable that target pest populations while preserving other organisms in the ecosystem, as far as is possible. Molecular biology holds particular promise as scientists continue to unravel the biochemistry of insect reproduction and development, and the chemical aspects of insect–plant interactions. Further development of microbial insecticides holds additional promise for useful tools in ecological pest management. Creative approaches to biocontrol, such as release of pesticide-resistant predators (Chapter 11), or release and conservation of parasitoids unspecific to the pest in question, are also worthy of further investigation. Hoy and Herzog (1985) discuss several situations in which this might be attempted.

In the future, fewer broad-spectrum chemical compounds will be available, even as the incidence of insecticide resistance increases somewhat as chemicals continue to be used for economy. Development of biocontrol and other narrow-spectrum techniques raises again the economic quandary faced by developers of insecticidal products in capitalistic economies: the need to maintain profits. Increased litigation and concomitant increases in liability insurance premiums for those involved in manufacture and use of broad-spectrum biocides may alter the uneven economic balance that prevails at present between broad- and narrow-spectrum methods. Another chemical accident on the order of that at Bhopal, India, in 1984 could well put a major international corporation out of business entirely.

Mathematical models and the computer simulations based on them present a fruitful starting point for developing alternative pest management systems, and both the principles and technology exist for developing realistic simulations of most arthropod pest populations. Additional intensive demographic studies, including detailed life tables, are necessary in order to refine and extend these models before they are widely implemented. Along with computer-based models will be increased use of artificial intelligence in pest management decision-making as the use of personal computers becomes widespread. Such programs are still in their infancy, yet hold great promise in pest management. Of course, it is essential that recommendations delivered via artificial intelligence be based on real intelligence; once again, there is no substitute for detailed and accurate information about the life systems of the pests involved (Chapters 3 through 5).

A critical problem facing those who would implement ecological pest management is how developing nations with limited resources are to handle the increasing costs of pest management in a climate of expanding human population and stable or declining agricultural profits. Ecological pest management obviously needs to be tailored to local conditions; it is not merely a matter of exporting western European or North American technology to the rest of the world, as has been the case with intensive chemical agriculture. More research is needed in such nations to render pest management cost-effective worldwide for long-term stability in agriculture.

Finally, there exists a need for additional education about real impacts of insect pests in both economic and aesthetic terms. Decisions regarding research and marketing priorities are made by a public (or its leadership) that is increasingly urban and isolated from both the outdoors generally and agricultural production systems in particular. Many North Americans seem to have an appalling lack of depth in understanding of how the natural world functions, and many become easy prey to unscrupulous panderers of misinformation about arthropods and pest control. It is hoped that the information presented in this book will help fill a few gaps in understanding. If it does that, the book will have been worth writing.

REFERENCES

Ables, J.R., R.P. West, and M. Shephard. 1975. Response of the housefly and its parasit-oids to Dimilin (TH-6040). J. Econ. Entomol. 68: 622–624.

Adkisson, P.L., and V.A. Dyck. 1980. Resistant varieties in pest management systems. pp. 233–251 in F.G. Maxwell and P.R. Jennings, eds. *Breeding plants resistant to insects*. Wiley, New York.

Allen, G.E., and J.E. Bath. 1980. The conceptual and institutional aspects of integrated pest management. BioScience 30: 658–664.

Alstad, D.N., and G.F. Edmunds, Jr. 1983. Adaptation, host specificity, and gene flow in the black pineleaf scale. pp. 413–426 in R.F. Denno and M.S. McClure, eds. *Variable plants and herbivores in natural and managed systems*. Academic Press, New York.

Altieri, M.A., and D.K. Letourneau. 1982. Vegetation management and biological con-trol in agroecosystems. Crop Prot. 1: 405–430.

Altieri, M.A., and L.L. Schmidt. 1986. Cover crops affect insect and spider populations in apple orchards. Calif. Agric. (Jan.–Feb.), pp. 15–17.

Altieri, M.A., and W.H. Whitcomb. 1979. The potential use of weeds in the manipula-tion of beneficial insects. Hort. Sci. 14: 12–18.

Altieri, M.A., A. van Schoonhoven, and J. Doll. 1977. The ecological role of weeds in integrated pest management systems: A review illustrated by bean (*Phaseolus vulgaris*) cropping systems. PANS 23: 195–205.

Altieri, M.A., C.A. Francis, A. van Schoonhoven, and J.D. Doll. 1978. A review of insect prevalence in maize (*Zea mays* L.) and bean (*Phaseolus vulgaris* L.) polycultural systems. Field Crops Res. 1: 33–49.

Altieri, M.A., P.B. Martin, and W.J. Lewis. 1983. A quest for ecologically based pest management systems. Envir. Manage. 7: 91–100.

Andaloro, J.T., C.W. Hoy, K.B. Rose, and A.M. Shelton. 1983. Evaluation of insecticide usage in the New York processing-cabbage pest management program. J. Econ. Entomol. 76: 1121–1124.

Anderson, R.M. 1982. Theoretical basis for the use of pathogens as biological control agents of pest species. Parasitol. 84: 3–33.

Anderson, R.M., and R.M. May. 1980. Infectious diseases and population cycles of forest insects. Science 210: 658–661.

Andrewartha, H.G., and L.C. Birch. 1954. *The distribution and abundance of animals*. University of Chicago Press, Chicago. 782 pp.

Apple, J.L., and R.F. Smith. 1976. *Insect pest management*. Plenum, New York. 200 pp.

Armbrust, E.J., and G.G. Gyrisco. 1982. Forage crops insect pest management. pp. 443–463 in R.L. Metcalf and W.L. Luckmann, eds. *Introduction to insect pest management*, 2nd ed. Wiley, New York.

Armbrust, E.J., B.C. Pass, D.W. Davis, R.G. Helgesen, G.R. Manglitz, R.L. Pienkowski, and C.G. Summers. 1980. General accomplishments toward better insect control in alfalfa. pp. 187–216 in C.B. Huffaker, ed. *New technology of pest control*. Wiley, New York.

Arnaud, P.H., Jr. 1978. *A host parasite catalogue of North American Tachinidae*. USDA SEA Misc. Publ. 1319. 860 pp.

Arthur, W. 1982. The evolutionary consequences of interspecific competition. Adv. Ecol. Res. 12: 127–187.

Askew, R.R. 1971. *Parasitic insects*. American Elsevier, New York. 316 pp.

Asmen, S.M., P.T. McDonald, and T. Prout. 1981. Field studies of genetic control systems for mosquitoes. Ann. Rev. Entomol. 26: 289–318.

Bach, C.E. 1980. Effects of plant density and diversity on the population dynamics of a specialist herbivore, the striped cucumber beetle *Acalymma vittata* (Fab.). Ecology 61: 1515–1530.

Bach, C.E. 1981. Host plant growth form diversity: Effects on abundance and feeding preference of a specialist herbivore, *Acalymma vittata* (Coleoptera: Chrysomelidae). Oecologia 50: 370–375.

Back, C., J. Boisvert, J.O. Lacoursière, and G. Charpentier. 1985. High-dosage treatment of a Quebec stream with *Bacillus thuringiensis* serovar. *israelensis*: Efficacy against black fly larvae (Diptera: Simuliidae) and impact on non-target insects. Canad. Entomol. 117: 1523–1534.

Baldwin, I.T., and J.C. Schultz. 1983. Rapid changes in tree leaf chemistry induced by damage: Evidence for communication between plants. Science 221: 277–279.

Ball, H.J. 1981. Insecticide resistance—a practical assessment. Bull. Entomol. Soc. Amer. 27: 261–262.

Barclay, H., and M. MacKauer. 1980. The sterile insect release method for pest control: a density-dependent model. Envir. Entomol. 9: 810–817.

Barclay, H.J., I.S. Otvos, and A.J. Thomson. 1985. Models of periodic inundation of parasitoids for pest control. Canad. Entomol. 117: 705–716.

Barnes, D.K., C.H. Hanson, R.H. Ratcliffe, T.H. Busbice, J.A. Schillinger, G.R. Buss, W.V. Campbell, R.W. Hemlken, and C.C. Blickenstaff. 1970. *The development of Team alfalfa—a multiple pest resistant alfalfa with moderate resistance to the alfalfa weevil*. Crop Res. Agric. Res. Serv. USDA 34-15. 41 pp.

Baumhover, A.H. 1966. Eradication of the screwworm fly. J. Amer. Med. Assoc. 196: 240–248.

Beattie, A.J., D.E. Breedlove, and P.R. Ehrlich. 1973. The ecology of the pollinators and predators of *Frasera speciosa*. Ecology 54: 81–91.

Bechinski, E.J., and L.P. Pedigo. 1981. Population dynamics and development of sampling plans for *Orius insidiosus* and *Nabis* species on soybeans. Envir. Entomol. 10: 956–959.

Beckendorf, S.K., and M.A. Hoy. 1985. Genetic improvement of arthropod natural enemies through selection, hybridization or genetic engineering techniques. pp. 167–187 in M.A. Hoy and D.C. Herzog, eds. *Biological control in agricultural IPM systems*. Academic Press, New York.

Bedding, R.A. 1984. Large scale production, storage, and transport of the insect parasitic nematodes *Neoplectana* spp. and *Heterorhabditis* spp. Ann. Appl. Biol. 104: 117–120.

Begon, M. 1979. *Investigating animal abundance: Capture-recapture for biologists*. University Park Press, Baltimore. 104 pp.

Bellinger, R.C., G.P. Dively II, and L.W. Douglass. 1981. Spatial distribution and sequential sampling of Mexican bean beetle defoliation on soybeans. Envir. Entomol. 10: 835–841.

Bennett, G.W., and J.M. Owens, eds. 1986. *Advances in urban pest management.* Van Nostrand Reinhold, New York. 397 pp.

Bennett, G.W., E.S. Runstrom, and J.A. Wieland. 1983. Pesticide use in homes. Bull. Entomol. Soc. Amer. 29(1): 31–38.

Berardi, G. 1983. Pesticide use in Italian food production. BioScience 33: 502–506.

Berenbaum, M. 1978. Toxicity of a furanocoumarin to armyworms: A case of biosynthetic escape from insect herbivores. Science 201: 532–533.

Berenbaum, M. 1983. Coumarins and caterpillars: A case for coevolution. Evolution 37: 163–179.

Bergman, J.M., and W.M. Tingey. 1979. Aspects of interaction between plant genotypes and biological control. Bull. Entomol. Soc. Amer. 25: 275–280.

Bernard, R.L. 1971. Two major genes for time of flowering and maturity in soybeans. Crop Sci. 11: 242–244.

Bernays, E.A., and S. Woodhead. 1982. Plant phenols utilized as nutrients by a phytophagous insect. Science 216: 201–203.

Beroza, M. 1972. Attractants and repellents for insect pest control. pp. 226–253 in National Academy of Sciences, USA. *Pest control: Strategies for the future.* Nat. Acad. Sci., Washington, DC.

Berryman, A.A. 1967. Estimation of *Dendroctonus brevicomis* mortality caused by insect predators. Canad. Entomol. 99: 1009–1014.

Berryman, A.A. 1981. *Population systems: A general introduction.* Plenum, New York. 222 pp.

Berryman, A.A. 1982. Biological control, thresholds, and pest outbreaks. Envir. Entomol. 11: 544–549.

Berryman, A.A., and R.W. Stark. 1985. Assessing the risk of forest insect outbreaks. Z. Angew. Entomol. 99: 199–208.

Bishop, O.N. 1967. *Statistics for biology.* Houghton–Mifflin, Boston. 182 pp.

Blomquist, G. 1979. The value of life-saving: Implications of consumption activity. J. Polit. Econ. 87: 540–558.

Blus, L.J., R.G. Heath, C.D. Gish, A.A. Belisle, and R.M. Prouty. 1971. Eggshell thinning in the brown pelican: implication of DDE. BioScience 21: 1213–1215.

Boldt, P.E., and J.J. Drea. 1980. Packaging and shipping beneficial insects for biological control. FAO Plant Prot. Bull. 28: 64–71.

Boucher, D.H., S. James, and K.H. Keeler. 1982. The ecology of mutualism. Ann. Rev. Ecol. Syst. 12: 315–347.

Bowers, W.S. 1981. How anti-juvenile hormones work. Amer. Zool. 21: 737–742.

Bowers, W.S., and R. Nishida. 1980. Juvicimenes: Potent juvenile hormone mimics from sweet basil. Science 209: 1030–1032.

Bowers, W.S., L.R. Nault, R.E. Webb, and S.R. Dutky. 1972. Aphid alarm pheromone: Isolation, identification, synthesis. Science 177: 1121–1122.

Brady, N.C. 1982. Chemistry and world food supplies. Science 218: 847–853.

Brower, J.H., and H.C. Scott. 1972. Gamma radiation sensitivity of the spider beetle, *Gibbium psylloides* (Coleoptera: Ptinidae). Canad. Entomol. 104: 1551–1556.

Brown, A.W.A. 1958. The spread of insecticide resistance in pest species. Adv. Pest Control Res. 2: 351–414.

Brown, A.W.A. 1968. Insecticide resistance comes of age. Bull. Entomol. Soc. Amer. 14: 3–9.

Brown, A.W.A. 1978. *Ecology of pesticides.* Wiley-Interscience, New York. 525 pp.

Brown, G.C., A.R. Lutgardo, and S.H. Gage. 1980. Data base management systems in IPM programs. Envir. Entomol. 9: 475-482.

Brown, M.W., and E.A. Cameron. 1982. Spatial distribution of adults of *Ooencyrtus kuvanae* (Hymenoptera: Encyrtidae) an egg parasite of *Lymantria dispar* (Lepidoptera: Lymantriidae). Canad. Entomol. 114: 1109-1120.

Brown, T.M., and W.G. Brogdon. 1987. Improved detection of insecticide resistance through conventional and molecular techniques. Ann. Rev. Entomol. 32: 145-162.

Brown, T.M., and A.W.A. Brown. 1974. Experimental induction of resistance to a juvenile hormone mimic. J. Econ. Entomol. 67: 799-801.

Brown, W.L., Jr. 1968. An hypothesis concerning the function of the metapleural glands in ants. Amer. Nat. 102: 188-191.

Brown, W.L., Jr., T. Eisner, and R.H. Whittaker. 1970. Allomones and kairomones: Transspecific chemical messengers. BioScience 20: 21-22.

Bucher, G.E., and H.H. Cheng. 1970. Use of trap plants for attracting cutworm larvae. Canad. Entomol. 102: 797-798.

Buckner, C.H., and J.C. Cunningham. 1972. The effect of the poxvirus of the spruce budworm, *Choristoneura fumiferana* (Lepidoptera: Tortricidae) on mammals and birds. Canad. Entomol. 104: 1333-1342.

Bundy, J.W. 1977. Consumer expectation and innovation in processed food production. pp. 227-235 in J.M. Cherrett and G.R. Sagar, eds. *Origin of pest, parasite, disease, and weed problems*. Blackwell, London.

Burdon, J.J., and D.R. Marshall. 1981. Biological control and the reproductive mode of weeds. J. Appl. Ecol. 18: 649-658.

Burrows, T.M., V. Sevacherian, H. Browning, and J. Baritelle. 1982. History and cost of the pink bollworm (Lepidoptera: Gelechiidae) in the Imperial Valley. Bull. Entomol. Soc. Amer. 28: 286-290.

Bush, G.L. 1969. Sympatric host race formation and speciation in frugivorous flies of the genus *Rhagoletis*. Evolution 23: 237-251.

Bush, G.L. 1975. Sympatric speciation in phytophagous parasitic insects. pp. 187-206 in P.W. Price, ed. *Evolutionary strategies of parasitic insects and mites*. Plenum, New York.

Bush, G.L., R.W. Neck, and G.B. Kitto. 1976. Screwworm eradication: inadvertent selection for noncompetitive ecotypes during mass rearing. Science 193: 491-493.

Byrne, D.N., E.H. Carpenter, E.M. Thomas, and S.T. Cotty. 1984. Public attitudes toward urban arthropods. Bull. Entomol. Soc. Amer. 30(2): 40-44.

Cain, M.L., J. Eccleston, and P.M. Kareiva. 1985. The influence of food plant dispersion on caterpillar searching success. Ecol. Entomol. 10: 1-7.

Calkins, C.O. 1983. Research on exotic insects. pp. 321-359 in C.L. Wilson and C.L. Graham, eds. *Exotic plant pests and North American agriculture*. Academic Press, New York.

Callahan, R.A., F.R. Holbrook, and F.R. Shaw. 1966. A comparison of sweeping and vacuum collecting certain insects affecting forage crops. J. Econ. Entomol. 59: 478-479.

Caltagirone, L.E. 1981. Landmark examples of classical biological control. Ann. Rev. Entomol. 26: 213-232.

Campbell, B.C., and S.S. Duffey. 1979. Tomatine and parasitic wasps: Potential incompatibility of plant antibiosis with biological control. Science 205: 700-702.

Cancelado, R.E., and E.B. Radcliffe. 1979. Action thresholds for green peach aphid on potatoes in Minnesota J. Econ. Entomol. 72: 606-609.

Cantelo, W.W., and L.L. Sanford. 1984. Insect population response to mixed and uniform plantings of resistant and susceptible plant material. Envir. Entomol. 13: 1443–1445.

Cantwell, G.E., ed. 1974. *Insect diseases*, vols. I and II. Marcel Dekker, New York. 595 pp.

Carson, R. 1962. *Silent Spring*. Houghton–Mifflin, Boston. 368 pp.

Carter, N., and A.F.G. Dixon. 1981. The "natural enemy ravine" in cereal aphid population dynamics: A consequence of predator activity or aphid biology? J. Anim. Ecol. 50: 605–611.

Carter, W. 1973. *Insects in relation to plant disease*, 2nd ed. Wiley, New York. 759 pp.

Cate, J.R. 1985. Cotton: status and current limitations to biological control in Texas and Arkansas. pp. 537–556 in M.A. Hoy and D.C. Herzog, eds. *Biological control in agricultural IPM systems*. Academic Press, New York.

Caughley, G., and J.H. Lawton. 1981. Plant herbivore systems. pp. 132–166 in R.M. May, ed. *Theoretical ecology: Principles and practice*. Blackwell, London.

Cave, R.D., and A.P. Gutierrez. 1983. *Lygus hesperus* field life table studies in cotton and alfalfa (Heteroptera: Miridae). Canad. Entomol. 115: 649–654.

Chesson, P.W., and W.W. Murdoch. 1986. Aggregation of risk: Relationships among host–parasitoid models. Amer. Nat. 127: 696–715.

Chew, F.S. 1981. Coexistence and local extinction in two pierid butterflies. Amer. Nat. 118: 655–672.

Chew, F.S., and J.E. Rodman. 1979. Plant resources for chemical defense. pp. 271–307 in G.A. Rosenthal and D.H. Janzen, eds. *Herbivores: Their interaction with secondary plant metabolites*. Academic Press, New York.

Chiverton, P.A. 1984. Pitfall-trap catches of the carabid beetle *Pterostichus melanarius* in relation to gut contents and prey densities, in insecticide treated and untreated spring barley. Entomol. Exp. Appl. 36: 23–30.

Chock, A. K. 1983. International cooperation on controlling exotic pests. pp. 479–498 in C.L. Wilson and C.L. Graham, eds. *Exotic plant pests and North American agriculture*. Academic Press, New York.

Claridge, M.F., and J. den Hollander. 1983. The biotype concept and its application to insect pests of agriculture. Crop Prot. 2: 85–95.

Clark, L.R., R.D. Hughes, P.W. Geier, and R.F. Morris. 1967. *The ecology of insect populations in theory and practice*. Methuen, London. 282 pp.

Clausen, C.P. 1951. The time factor in biological control. J. Econ. Entomol. 44: 1–9.

Clausen, C.P. 1956. *Biological control of insect pests in the continental United States*. USDA Tech. Bull. 1139. 151 pp.

Clausen, C.P., ed. 1977. *Introduced parasites and predators of arthropod pests and weeds: A world review*. USDA Agr. Handbook 480. 551 pp.

Cocke, J.R., Jr. 1981. *New advances against the screwworm*. Texas A&M Exp. Sta. Bull. L-1089. 12 pp.

Cole, H., D. MacKenzie, C.B. Smith, and E.L. Bergman. 1968. Influence of various persistent chlorinated insecticides on the macro and micro element constituents of *Zea mays* and *Phaseolus vulgaris* growing in soil containing various amounts of these materials. Bull. Envir. Contam. Toxicol. 3: 141–154.

Collins, F.H., R.K. Sakai, K.D. Vernick, S. Paskewitz, D.C. Seeley, L.H. Miller, W.E. Collins, C.C. Campbell, and R.W. Gwadz. 1986. Genetic selection of a *Plasmodium*-refractory strain of the malaria vector *Anopheles gambiae*. Science 234: 607–610.

Cooper, R.M. 1981. A field and laboratory evaluation of Bay SIR 8514 for control of the Colorado potato beetle, *Leptinotarsa decimlineata* (Say), on potatoes. MS thesis, Ohio State University. 80 pp.

Cornell, H., and D. Pimentel. 1978. Switching in the parasitoid *Nasonia vitripennis* and its effects on host competition. Ecology 59: 297-308.

Cothran, W.R., and C.G. Summers. 1972. Sampling for the Egyptian alfalfa weevil: a comment on the sweep net method. J. Econ. Entomol. 65: 689-691.

Coulman, G.A., S.R. Reice, and R.L. Tummala. 1972. Population modelling: A systems approach. Science 171: 518-521.

Coulson, R.N., and M.C. Saunders. 1987. Computer-assisted decision-making as applied to entomology. Ann. Rev. Entomol. 32: 415-437.

Coulson, R.N., and J.A. Witter. 1984. *Forest entomology ecology and management.* Wiley, New York. 669 pp.

Coulson, R.N., A.M. Mayyasi, J.L. Folt, and F.P. Hain. 1976. Interspecific competition between *Monochamus titillator* and *Dendroctonus frontalis*. Envir. Entomol. 5: 235-247.

Coulson, R.N., W.A. Leuschner, J.L. Foltz, P.E. Pulley, F.P. Hain, and T.L. Payne. 1980. Approach to research and forest pest management for southern pine beetle control. pp. 449-469 in C.B. Huffaker, ed. *New technology of pest control.* Wiley, New York.

Croft, B.A., and S.A. Hoying. 1977. Competitive displacement of *Panonychus ulmi* (Acari: Tetranychidae) by *Aculus schlechtendali* (Acari: Eriophyidae) in apple orchards. Canad. Entomol. 109: 1025-1034.

Croft, B.A., J.L. Howes, and S.M. Welch. 1976. A computer-based, extension pest management delivery system. Envir. Entomol. 5: 20-34.

Crooks, E., K. Havel, M. Shannon, G. Snyder, and T. Wallenmaier. 1983. Stopping pest introductions. pp. 239-259 in C.L. Wilson and C.L. Graham, eds. *Exotic plant pests and North American agriculture.* Academic Press, New York.

Curry, G.L., P.J. Sharpe, D.W. DiMichele, and J.R. Cate. 1980. Towards a management model of the cotton-boll weevil ecosystem. J. Envir. Management 11: 187-223.

DaCosta, C.P., and C.M. Jones. 1971. Cucumber beetle resistance and mite susceptibility controlled by the bitter gene in *Cucumis sativus*. Science 172: 1145-1146.

Dahlsten, D.L., and W.A. Copper. 1979. The use of nesting boxes to study the biology of the mountain chickadee (*Parus gambeli*) and its impact on selected forest insects. pp. 217-260 in J.G. Dickson, R.N. Connor, R.R. Fleet, J.A. Jackson, and J.C. Kroll. *The role of insectivorous birds in forest ecosystems.* Academic Press, New York.

Dame, D.A., and C.H. Schmidt. 1970. The sterile male technique against tsetse flies, *Glossina* spp. Bull. Entomol. Soc. Amer. 16: 24-30.

Danthanarayana, W. 1983. Population ecology of the light brown apple moth, *Epiphyas postvittana* (Lepidoptera: Tortricidae). J. Anim. Ecol. 52: 1-33.

Davidson, R.H., and W.F. Lyon. 1979. *Insect pests of farm, garden and orchard,* 7th ed. Wiley, New York. 596 pp.

DeBach, P. 1947. Cottony cushion scale, vedelia, and DDT in central California. Calif. Citrogr. 32: 406-407.

DeBach, P.H., ed. 1964. *Biological control of insect pests and weeds.* Reinhold, New York. 844 pp.

DeBach, P., and R.A. Sundby. 1963. Competitive displacement between ecological homologues. Hilgardia 35: 105-166.

DeBach, P., E.J. Dietrick, and C.A. Fleschner. 1949. A new technique for evaluating the efficiency of entomophagous insects in the field. J. Econ. Entomol. 42: 546.

DeBach, P., C.A. Fleschner, and E.J. Dietrick. 1951. A biological check method for evaluating the effectiveness of entomophagous insects. J. Econ. Entomol. 44: 763-766.

DeBarjac, H. 1978. Un nouveau candidat à la lutte biologique contre les moustiques: *Bacillus thuringiensis* var. *israelensis*. Entomophaga 23: 309-319.

Deevey, E.S. 1947. Life tables for natural populations of animals. Quart. Rev. Biol. 22: 283-314.

DeLong, D.M. 1932. Some problems encountered in the estimation of insect populations by the sweeping method. Ann. Entomol. Soc. Amer. 25: 13-17.

Dempster, J.P. 1983. The natural control of populations of butterflies and moths. Biol. Rev. 58: 461-481.

Denlinger, D.L. 1981. The physiology of pupal diapause in Diptera. Entomol. Gen. 7: 245-259.

Denno, R.F. 1977. Comparison of the assemblages of sap-feeding insects (Homoptera-Hemiptera) inhabiting two structurally different marsh grasses in the genus *Spartina*. Envir. Entomol. 6: 359-372.

Denno, R.F. 1979. The relation between habitat stability and the migration tactics of planthoppers. Misc. Publ. Entomol. Soc. Amer. 11: 41-49.

Denno, R.F. 1983. Tracking variable host plants in space and time. pp. 291-341 in R.F. Denno and M.S. McClure, eds. *Variable plants and herbivores in natural and managed systems*. Academic Press, New York.

Denno, R.F. 1985. Fitness, population dynamics and migration in planthoppers: The role of host plants. Univ. Texas Contrib. Marine Sci. Suppl. 27: 623-640.

Denno, R.F., and M.S. McClure, eds. 1983. *Variable plants and herbivores in natural and managed systems*. Academic Press, New York. 717 pp.

Dethier, V.G. 1976. *Man's plague? Insects and agriculture*. Darwin Press, Princeton, NJ. 237 pp.

Dickson, J.G., R.N. Connor, R.R. Fleet, J.A. Jackson, and J.C. Kroll. 1979. *The role of insectivorous birds in forest ecosystems*. Academic Press, New York. 381 pp.

Dietrick, E.J. 1961. An improved backpack motor fan for suction sampling of insect populations. J. Econ. Entomol. 54: 394-395.

Doane, C.C., and M.L. McManus. 1981. *The gypsy moth: research toward integrated pest management*. USDA Tech. Bull. 1584. 757 pp.

Dolinger, P.M., P.R. Ehrlich, W.L. Fitch, and D.E. Breedlove. 1973. Alkaloid and predation patterns in Colorado lupine populations. Oecologia 13: 191-204.

Dondale, C.D. 1972. Effects of carbofuran on arthropod populations and crop yield in hayfields. Canad. Entomol. 104: 1433-1437.

Donley, D.L. 1976. Insect impact on production of oak timber. South. Lumberman (Dec.). 4 pp.

Dover, M., and B. Croft. 1984. *Getting tough: Public policy and the management of insecticide resistance*. Study 1, World Resources Inst., Washington. 80 pp.

Dover, M.J., and B.A. Croft. 1986. Pesticide resistance and public policy. BioScience 36: 78-85.

Dowell, R.V., R.H. Cherry, and G.E. Fitzpatrick. 1979. Citrus pests in an urban environment. Fla. Sci. 42: 196-200.

Drea, J.J., Jr., and R.M. Hendrickson. 1986. Analysis of a successful classical biological control project: the alfalfa blotch leafminer (Diptera: Agromyzidae) in the northeastern United States. Envir. Entomol. 15: 448-455.

Drooz, A.T., A.E. Bustillo, G.F. Fedde, and V.H. Fedde. 1977. North American egg parasite successfully controls a different host genus in South America. Science 197: 390-391.

Dunn, Z.A., and D.P.H. Kempton. 1972. Resistance to attack by *Brevicoryne brassicae* among plants of Brussels sprouts. Ann. Appl. Biol. 72: 1-11.

Dysart, J.R., and W.H. Day. 1976. *Release and recovery of introduced parasites of the alfalfa weevil in eastern North America.* Agr. Res. Serv. USDA Prod. Res. Rep. 167. 61 pp.

Edwards, P.J., and S.D. Wratten. 1985. Induced plant defenses in insect grazing: Fact or artifact? Oikos 44: 70–74.

Ehler, L.E. 1976. The relationship between theory and practice in biological control. Bull. Entomol. Soc. Amer. 22: 319–321.

Ehler, L.E. 1978. Some aspects of urban agriculture. pp. 349–357 in G.W. Frankie and C.S. Koehler, eds. *Perspectives in urban entomology.* Academic Press, New York.

Ehler, L.E. 1982. Foreign exploration in California. Envir. Entomol. 11: 525–530.

Ehler, L.E. 1985. Species-dependent mortality in a parasite guild and its relevance to biological control. Envir. Entomol. 14: 1–6.

Ehler, L.E., and P.C. Endicott. 1984. Effect of malathion-bait sprays on biological control of insect pests of olive, citrus, and walnut. Hilgardia 52: 1–47.

Ehler, L.E., and R.W. Hall. 1982. Evidence for competitive exclusion of introduced natural enemies in biological control. Envir. Entomol. 11: 1–4.

Ehler, L.E., and J.C. Miller. 1978. Biological control in temporary agroecosystems. Entomophaga 23: 207–212.

Ehler, L.E., P.C. Endicott, M.B. Hertlein, and B. Alvarado-Rodriguez. 1984. Medfly eradication in California: impact of malathion-bait sprays on an endemic gall midge and its parasitoids. Entomol. Exp. Appl. 36: 201–208.

Ehrlich, P.R., and L.C. Birch. 1967. The "balance of nature" and "population control." Amer. Nat. 101: 97–107.

Ehrlich, P.R., and P.H. Raven. 1965. Butterflies and plants: A study in coevolution. Evolution 18: 586–608.

Ehrlich, P.R., R.R. White, M.C. Singer, S.W. McKechnie, and L.I. Gilbert. 1975. Checkerspot butterflies: A historical perspective. Science 188: 221–228.

Eichers, T.R. 1981. *Farm pesticide economic evaluation, 1981.* USDA Econ. Res. Serv. Agr. Econ. Report #464. 21 pp.

Elhag, E.A. 1976. Evaluation of feeding of the potato leafhopper, *Empoasca fabae* (Harris), on growth and nitrogen content of alfalfa. MS thesis, Ohio State University, Columbus. 46 pp.

Elkinton, J.S., and R.T. Cardé. 1980. Distribution, dispersal, and apparent survival of male gypsy moths as determined by capture in pheromone-baited traps. Envir. Entomol. 9: 729–737.

Embree, D.G. 1966. The role of introduced parasites in the control of the winter moth in Nova Scotia. Canad. Entomol. 98: 1159–1168.

Embree, D.G. 1979. The ecology of colonizing species, with special emphasis on animal invaders. pp. 51–65 in D.J. Horn, R. Mitchell, and G.R. Stairs, eds. *Analysis of ecological systems.* Ohio State University Press, Columbus.

Erwin, T.L. 1983. Tropical forest canopies: The last biotic frontier. Bull. Entomol. Soc. Amer. 29 (1): 14–19.

Everett, T.R., H.C. Chiang, and T.R. Hibbs. 1958. Some factors influencing populations of European corn borer (*Pyrausta nubilalis* (Hbn.)) in the North Central states. Minn. Agr. Exp. Sta. Tech. Bull 229, 63 pp.

Faeth, S.H., and D. Simberloff. 1981. Population regulation of a leafmining insect, *Cameraria* sp. nov., at increased field densities. Ecology 62: 620–624.

Falcon, L.A. 1985. Development and use of microbial insecticides. pp. 229–242 in M.A. Hoy and D.C. Herzog, eds. *Biological control in agricultural IPM systems.* Academic Press, New York.

Faulkner, P., and D.G. Boucias. 1985. Genetic improvement of insect pathogens: Emphasis on the use of baculoviruses. pp. 263–281 in M.A. Hoy and D.C. Herzog, eds. *Biological control in agricultural IPM systems.* Academic Press, New York.

Feder, G. 1979. Pesticides, information, and pest management under uncertainty. Amer. J. Agr. Econ. 55: 198–201.

Feeny, P.P. 1970. Seasonal changes in oak leaf tannins and nutrients as a cause of spring feeding by winter moth caterpillars. Ecology 51: 565–581.

Feeny, P.P. 1975. Biochemical coevolution between plants and their insect herbivores. pp. 3–19 in L.E. Gilbert and P.R. Raven, eds. *Coevolution of animals and plants.* University of Texas Press, Austin.

Feng, G.K., Y.L. Chou, K.S. Chang, and S.H. Nieh. 1977. Studies on control of European corn borers by using trichogrammatid egg parasites. Acta Entomol. Sinica 20: 253–259.

Fenton, F.A., and D.E. Howell. 1957. A comparison of five methods of sampling alfalfa fields for arthropod pests. Ann. Entomol. Soc. Amer. 50: 606–611.

Finch, S., G. Skinner, and G.H. Freeman. 1975. The distribution and analysis of cabbage root fly egg populations. Ann. Appl. Biol. 79: 1–18.

Fiori, B.J., and D.D. Dolan. 1981. Field tests for *Medicago* resistance against potato leafhopper (Homoptera: Cicadellidae). Canad. Entomol. 113: 1049–1053.

Fleischer, S.J., W.A. Allen, J.M. Luna, and R.L. Peinkowski. 1982. Absolute-density estimation from sweep sampling, with a comparison of absolute density sampling techniques for adult potato leafhopper in alfalfa. J. Econ. Entomol. 75: 425–430.

Flinn, P.W., R.A.J. Taylor, and A.A. Hower. 1986. Predictive model for the population dynamics of potato leafhopper, *Empoasca fabae* (Homoptera: Cicadellidae) on alfalfa. Envir. Entomol. 15: 898–904.

Flint, M.L., and R. van den Bosch. 1981. *Introduction to integrated pest management.* Plenum, New York. 240 pp.

Follett, P.A., B.A. Croft, and P.H. Westigard. 1985. Regional resistance to insecticides in *Psylla pyricola* from pear orchards in Oregon. Canad. Entomol. 117: 565–573.

Force, D.C. 1972. r- and K-strategists in endemic host-parasite communities. Bull. Entomol. Soc. Amer. 18: 135–137.

Force, D.C. 1974. Ecology of insect host-parasite communities. Science 184: 624–632.

Foster, W.A., and K.I. Lutes. 1985. Tests of ultrasonic emissions on mosquito attraction to hosts in a flight chamber. J. Amer. Mosq. Contr. Assoc. 1: 199–202.

Fowler, S.V., and J.H. Lawton. 1985. Rapidly induced defenses and talking trees: The devil's advocate position. Amer. Nat. 126: 181–195.

Fox, L.R., and P.A. Morrow. 1981. Specialization: Species property or local phenomenon? Science 211: 887–893.

Fraenkel, G.S. 1959. The raison d'être of secondary plant substances. Science 129: 1466–1470.

Frankie, G.W., and C.S. Koehler, eds. 1983. *Urban entomology: Interdisciplinary perspectives.* Praeger, New York. 493 pp.

Frisbie, R.E., and P.L. Adkisson. 1985. IPM: definitions and current status in U.S. agriculture. pp. 41–51 in M.A. Hoy and D.C. Herzog, eds. *Biological control in agricultural IPM systems.* Academic Press, New York.

Fritz, R.S. 1983. Ant protection of a host plant's defoliator: Consequence of an ant-membracid mutualism. Ecology 64: 789–797.

Fronk, W.D. 1985. Vegetable crop insects. pp. 371–398 in R.E. Pfadt, ed. *Fundamentals of applied entomology,* 4th ed. Macmillan, New York.

Fry, D.M., and C.K. Toone. 1981. DDT-induced feminization of gull embryos. Science 213: 923–924.

Führer, E. 1985. Basic problems of long-term regulation of insect pest populations in a managed forest. Z. Angew. Entomol. 99: 67–73.

Furniss, M.M., L.N. Kline, R.F. Schmitz, and J.A. Rudinsky. 1972. Tests of three pheromones to induce or disrupt aggregation of Douglas-fir beetles (Coleoptera: Scolytidae) on live trees. Ann. Entomol. Soc. Amer. 65: 1227–1232.

Furniss, M.M., G.E. Daterman, L.N. Kline, M.D. McGregor, G.C. Trostle, L.F. Dettiger, and J.A. Rudinsky. 1974. Effectiveness of the Douglas fir beetle anti-aggregation pheromone, methylcyclohexone, at three concentrations and spacings around felled trees. Canad. Entomol. 106: 381–392.

Futuyma, D.J. 1976. Food plant specialization and environmental predictability in Lepidoptera. Amer. Nat. 110: 285–292.

Futuyma, D.J., and F. Gould. 1979. Associations of plants and insects in a deciduous forest. Ecol. Monog. 49: 33–50.

Futuyma, D.J., and S.S. Wasserman. 1980. Resource concentration and herbivory in oak forests. Science 210: 920–922.

Fuxa, J.R. 1987. Ecological considerations for the use of entomopathogens in IPM. Ann. Rev. Entomol. 32: 225–251.

Fye, R.E. 1983. Impact of volcanic ash on pear psylla (Homoptera: Psyllidae) and associated predators. Envir. Entomol. 12: 222–226.

Gage, S.H., M.E. Whalon, and D.J. Miller. 1982. Pest event scheduling systems for biological monitoring and pest management. Envir. Entomol. 11: 1127–1133.

Gall, G.A.E., J.J. Cech, Jr., R. Garcia, V.H. Resh, and R.K. Washino. 1980. Mosquito fish—an established predator. Calif. Agr. 34: 21–22.

Gallun, R.L., and G.S. Khush. 1980. Genetic factors affecting expression and stability of resistance. pp. 63–85 in F.G. Maxwell and P.R. Jennings, eds. *Breeding plants resistant to insects*. Wiley, New York.

Gallun, R.L., K.J. Starks, and W.D. Guthrie. 1975. Plant resistance to insects attacking cereals. Ann. Rev. Entomol. 20: 337–351.

Gaston, L.K., R.S. Kane, H.H. Shorey, and D. Sellers. 1977. Controlling the pink bollworm by disrupting sex pheromone communication between adult moths. Science 196: 904–905.

Gebhardt, M.R., T.C. Daniel, E.E. Schweizer, and R.R. Allmaras. 1985. Conservation tillage. Science 230: 625–630.

Georghiou, G.P., and T. Saito, eds. 1983. *Pest resistance to pesticides*. Plenum, New York. 809 pp.

Georghiou, G.P., and C.E. Taylor. 1977. Genetic and biological influences in the evolution of insecticide resistance. J. Econ. Entomol. 70: 319–323.

Getz, W.M., and A.P. Gutierrez. 1982. A perspective on systems analysis in crop production and integrated pest management. Ann. Rev. Entomol. 27: 447–466.

Gibson, W.P., and R.C. Berberet. 1974. Histological studies on encapsulation of *Bathyplectes curculionis* eggs by larvae of the alfalfa weevil. Ann. Entomol. Soc. Amer. 67: 588–590.

Goeden, R.D., and L.T. Kok. 1986. Comments on a proposed "new" approach for selecting agents for the biological control of weeds. Canad. Entomol. 118: 51–58.

Goodman, D. 1975. The theory of diversity–stability relationships in ecology. Quart. Rev. Biol. 50: 237–266.

Gould, F. 1984. Role of behavior in the evolution of insect adaptation to insecticides and resistant host plants. Bull. Entomol. Soc. Amer. 30(4): 34–41.

Gould, W.P., and R.L. Jeanne. 1984. *Polistes* wasps (Hymenoptera: Vespidae) as control agents for lepidopterous cabbage pests. Envir. Entomol. 13: 150–156.

Greany, P.D., S.B. Vinson, and W.J. Lewis. 1984. Insect parasitoids: Finding new opportunities for biological control. BioScience 34: 690-696.

Green, R.H. 1970. On fixed precision level sequential sampling. Res. Pop. Ecol. 12: 249-251.

Greene, C.R., E.G. Rajotte, G.W. Norton, R.A. Kramer, and R.M. McPherson. 1985. Revenue and risk analysis of soybean pest management options in Virginia. J. Econ. Entomol. 78: 10-18.

Grier, J.W. 1982. Ban of DDT and subsequent recovery of reproduction in bald eagles. Science 218: 1232-1235.

Gross, H.P., W.F. Baldwin, and A.S. West. 1972. Introductory studies on the use of radiation in the control of black flies (Diptera: Simuliidae). Canad. Entomol. 104: 1217-1222.

Grothusen, J. 1984. A simple calculator program for flexible sequential sampling of insects. Bull. Entomol. Soc. Amer. 30 (3): 35-37.

Gruys, P. 1982. Hits and misses: The ecological approach to pest control in orchards. Entomol. Exp. Appl. 31: 70-87.

Guppy, J.C., and D.G. Harcourt. 1970. Spatial pattern of the immature stages and teneral adults of *Phyllophaga* species (Coleoptera: Scarabaeidae) in a permanent meadow. Canad. Entomol. 102: 1354-1359.

Guppy, J.C., and D.G. Harcourt. 1973. A sampling plan for studies on the population dynamics of white grubs, *Phyllophaga* species. Canad. Entomol. 105: 479-483.

Guthrie, F.E., R.L. Rabb, T.G. Bowery, F.R. Lawson, and R.L. Baron. 1959. Control of hornworms and budworms on tobacco with reduced insecticide dosage. Tobacco Sci. 3: 65-68.

Gutierrez, A.P., J.R. Christensen, W.B. Loew, C.M. Merritt, C.G. Summers, and W.R. Cothran. 1976. Alfalfa and the Egyptian alfalfa weevil (Coleoptera: Curculionidae). Canad. Entomol. 108: 635-648.

Gutierrez, A.P., T.F. Leigh, Y. Wang, and R.D. Cave. 1977a. An analysis of cotton production in California: *Lygus hesperus* (Heteroptera: Miridae) injury—an evaluation. Canad. Entomol. 109: 1375-1386.

Gutierrez, A.P., G.D. Butler, Y. Wang, and D. Westphal. 1977b. The interaction of pink bollworm (Lepidoptera: Gelechiidae), cotton, and weather: A detailed model. Canad. Entomol. 109: 1457-1468.

Gutierrez, A.P., D.W. DeMichele, Y. Wang, G.L. Curry, R. Skeith, and L.G. Brown. 1980. The systems approach to research and decision making for cotton pest control. pp. 155-186 in C.B. Huffaker, ed. *New technology of pest control*. Wiley, New York.

Gutierrez, A.P., J.U. Baumgaertner, and C.G. Summers. 1984. Multitrophic models of predator-prey energetics. Canad. Entomol. 116: 923-963.

Hagen, K.S., E.F. Sewall, Jr., and R.L. Tassan. 1970. The use of food sprays to increase the effectiveness of entomophagous insects. Proc. Tall Timbers Conf. Ecol. Anim. Contr. Hab. Man. 2: 59-82.

Hall, R.A., and B. Papierok. 1982. Fungi as biological control agents of arthropods of agricultural and medial importance. Parasitol. 84: 205-240.

Hall, R.W., L.E. Ehler, and B. Bisabri-Ershadi. 1980. Rate of success in classical biological control of arthropods. Bull. Entomol. Soc. Amer. 26: 111-114.

Hammond, R.B., and L.P. Pedigo. 1982. Determination of yield-loss relationships for two soybean defoliators by using simulated insect-defoliation techniques. J. Econ. Entomol. 75: 102-107.

Hanover, J.W. 1980. Breeding forest trees resistant to insects. pp. 487-511 in F.G.

Maxwell and P.R. Jennings, eds. *Breeding plants resistant to insects.* Wiley, New York.

Harcourt, D.G. 1963. Population dynamics of *Leptinotarsa decimlineata* (Say) in eastern Ontario. I. Spatial pattern and transformation of field counts. Canad. Entomol. 95: 813–820.

Harcourt, D.G. 1964. Population dynamics of *Leptinotarsa decimlineata* (Say) in eastern Ontario. II. Population and mortality estimation during six age intervals. Canad. Entomol. 96: 1190–1198.

Harcourt, D.G. 1969. The development and use of life tables in the study of natural insect populations. Ann. Rev. Entomol. 14: 175–196.

Harcourt, D.G. 1970. Crop life tables as a pest management tool. Canad. Entomol. 102: 950–955.

Harcourt, D.G. 1971. Population dynamics of *Leptinotarsa decimlineata* (Say) in eastern Ontario. III. Major population processes. Canad. Entomol. 103: 1049–1061.

Harcourt, D.G., and M.R. Binns. 1980. A sampling system for estimating egg and larval populations of *Agromyza frontella* (Diptera: Agromyzidae) in alfalfa. Canad. Entomol. 112: 375–385.

Harcourt, D.G., J.C. Guppy, and M.R. Binns. 1977. The analysis of intrageneration changes in eastern Ontario populations of the alfalfa weevil, *Hypera postica* (Coleoptera: Curculionidae). Canad. Entomol. 109: 1521–1534.

Harcourt, D.G., J.C. Guppy, and M.R. Binns. 1984. Analysis of numerical change in subeconomic populations of the alfalfa weevil, *Hypera postica* (Coleoptera: Curculionidae) in eastern Ontario. Envir. Entomol. 13: 1627–1633.

Hargrove, W.W., D.A. Crossley, Jr., and T.R. Seastedt. 1984. Shifts in insect herbivory in the canopy of black locust, *Robinia pseudoacacia*, after fertilization. Oikos 43: 322–328.

Harris, K.F., and K. Maramorosch, eds. 1982. *Pathogens, vectors, and plant diseases: Approaches to control.* Academic Press, New York. 310 pp.

Harrison, R.G. 1980. Dispersal polymorphisms in insects. Ann. Rev. Ecol. Syst. 11: 95–118.

Hassell, M.P. 1978. *The dynamics of arthropod predator–prey systems.* Princeton University Press, Princeton, NJ. 237 pp.

Hassell, M.P. 1985. Insect natural enemies as regulating factors. J. Anim. Ecol. 54: 323–334.

Hassell, M.P., J.H. Lawton, and J.R. Beddington. 1977. Sigmoid functional responses by invertebrate predators and parasitoids. J. Anim. Ecol. 46: 249–262.

Haukioja, E., and S. Neuvonen. 1985. Induced long-term resistance of birch foliage against defoliators: Defensive or incidental? Ecology 66: 1303–1308.

Haws, B.A. and D.W. Davis. 1985. Insects of legumes. pp. 310–338 in R.E. Pfadt, ed. *Fundamentals of applied entomology*, 4th ed. Macmillan, New York.

Headley, J.C. 1972. Economics of agricultural pest control. Ann. Rev. Entomol. 17: 273–286.

Headley, J.C. 1975. The economics of pest management. pp. 75–99 in R.L. Metcalf and W.H. Luckmann, eds. *Introduction to insect pest management.* Wiley, New York.

Headley, J.C., and M.A. Hoy. 1986. The economics of integrated mite management in almonds. Calif. Agr. (Jan.–Feb.) pp. 28–30.

Hearn, A.B., P.M. Ives, P.M. Room, N.J. Thomson, and L.T. Wilson. 1981. Computer-based cotton pest management in Australia. Field Crops Res. 4: 321–332.

Heimpel, A.M., E.D. Thomas, J.R. Adams, and L.J. Smith. 1973. The presence of nuclear polyhedrosis viruses of *Trichoplusia ni* on cabbage from the market shelf. Envir. Entomol. 2: 72–75.

Heinrich, B. 1979. Foraging strategies of caterpillars: Leaf damage and possible predator avoidance strategies. Oecologia 42: 325–337.

Heinrichs, E.A., W.H. Reissig, S. Valencia, and S. Chelliah. 1982. Rates and effect of resurgence-inducing insecticide on populations of *Nilaparvata lugens* (Homoptera: Delphacidae) and its predators. Envir. Entomol. 11: 1269–1273.

Hendry, L.B., J.K. Wichmann, D.M. Hindenlang, K.M. Weaver, and S.H. Korzenioski. 1976. Plants—the origin of kairomones utilized by parasitoids of phytophagous insects? J. Chem. Ecol. 2: 271–283.

Herbert, H.J. 1981. Biology, life tables, and intrinsic rate of increase of the european red mite, *Panonychus ulmi* (Acarina: Tetranychidae). Canad. Entomol. 113: 65–71.

Herzog, D.C., and J.E. Funderburk. 1985. Plant resistance and cultural practice interactions with biological control. pp. 67–88 in M.A. Hoy and D.C. Herzog, eds. *Biological control in agricultural IPM systems*. Academic Press, New York.

Heynen, C. 1985. Untersuchungen zum Einfluss von Diflubenzuron (Dimilin) auf das Wirt-Parasit-System *Spodoptera littoralis* Boisd. (Lepidoptera, Noctuidae)/*Microplitis rufiventris* Kok. (Hymenoptera, Braconidae). Z. Angew. Entomol. 100: 113–132.

Higgins, R.A., L.P. Pedigo, and D.W. Staniforth. 1984. Effect of velvetleaf, competition and defoliation simulating a green cloverworm (Lepidoptera: Nocturidae) outbreak in Iowa on indeterminate soybean yield components and economic decision levels. Envir. Entomol. 13: 917–925.

Higley, L.G., L.P. Pedigo, and K.R. Ostlie. 1986. DEGDAY: A program for calculating degree-days, and assumptions behind the degree-day approach. Envir. Entomol. 15: 999–1016.

Hill, R.E., and W.J. Gary. 1979. Effects of the microsporidian, *Nosema pyrausta*, on field populations of European corn borer in Nebraska. Envir. Entomol. 8: 91–95.

Hill, R.E., and Z.B. Mayo. 1980. Distribution and abundance of corn rootworm species as influenced by topography and crop rotation in eastern Nebraska. Envir. Entomol. 9: 122–127.

Hillebrandt, P.M. 1960. The economic theory of the use of pesticides. Part 1. The dosage-response curve, the rate of application and the area to be treated. J. Agr. Econ. 13: 464–472.

Hodek, I., ed. 1986. *Ecology of aphidophagous insects*. II. Academia, Prague and W. Junk, The Hague. 562 pp.

Hoffman, C.H., and E.P. Merkel. 1948. Fluctuations in insect populations associated with aerial applications of DDT. J. Econ. Entomol. 41: 464–473.

Hoffman, C.H., H.K. Townes, H.H. Swift, and R.I. Sailer. 1949. Field studies on the effects of airplane applications of DDT on forest invertebrates. Ecol. Monogr. 19: 1–46.

Hokkanen, H., and D. Pimentel. 1984. New approach for selecting biological control agents. Canad. Entomol. 116: 1109–1121.

Holdsworth, R.P., Jr. 1968. Integrated control: effect on European red mite and its more important predators. J. Econ. Entomol. 61: 1602–1607.

Holling, C.S. 1959. The components of predation as revealed by a study of predation of the European pine sawfly. Canad. Entomol. 91: 293–320.

Holmes, R.T., J.C. Schultz, and P. Nothnagle. 1979. Bird predation on forest insects: An exclosure experiment. Science 206: 462–463.

Hori, K. 1974. Plant growth-promoting factor in the salivary gland of the bug *Lygus disponsi*. J. Ins. Physiol. 20: 1623–1627.

Hori, K. 1976. Plant growth-regulating factor in the salivary glands of several heteropterous insects. Comp. Biochem. Physiol. B 53: 435–438.

Horn, D.J. 1971. The relationship between a parasite, *Tetrastichus incertus* (Hymenoptera: Eulophidae) and its host, the alfalfa weevil, *Hypera postica* (Coleoptera: Curculionidae) in New York. Canad. Entomol. 103: 83–94.

Horn, D.J. 1976. *Biology of insects*. Saunders, Philadelphia. 439 pp.

Horn, D.J. 1981. Effect of weedy backgrounds on colonization of collards by green peach aphid, *Myzus persicae*, and its major predators. Envir. Entomol. 10: 285–289.

Horn, D.J. 1983. Selective mortality of parasitoids and predators of *Myzus persicae* on collards treated with malathion, carbaryl, or *Bacillus thuringiensis*. Entomol. Exp. Appl. 34: 208–211.

Horn, D.J. 1984. Vegetational complexity and parasitism of green peach aphids (*Myzus persicae* (Sulzer) (Homoptera: Aphidae)) on collards. J. New York Entomol. Soc. 92: 19–26.

Horn, D.J. 1986. Aphid–parasitoid interactions: Influence of weed management on the fauna of collards. pp. 285–290 in I. Hodek, ed. *Ecology of aphidophagous insects*. Academia, Prague and W. Junk, The Hague.

Horn, D.J., and R.V. Dowell. 1979. Parasitoid ecology and biological control in ephemeral crops. pp. 281–307 in D.J. Horn, R. Mitchell, and G.R. Stairs, eds. *Analysis of ecological systems*. Ohio State University Press, Columbus.

Horn, D.J., and R.W. Wadleigh. 1987. Resistance of aphid natural enemies to insecticides. in P. Harrewijn and A.K. Minks, eds. *Aphids, their biology, natural enemies, and control*. Elsevier, Amsterdam (in press).

Houck, M.A., and R.E. Strauss. 1985. The comparative study of functional responses: Experimental design and statistical interpretation. Canad. Entomol. 117: 617–629.

Howard, F.W., and A.D. Oliver. 1978. Arthropod populations in permanent pastures treated and untreated with mirex for red imported fire ant control. Envir. Entomol. 7: 901–903.

Howarth, F.G. 1983. Classical biocontrol: Panacea or Pandora's box? Proc. Hawaiian Entomol. Soc. 24: 239–245.

Howe, W.L., and O.F. Smith. 1957. Resistance to the spotted alfalfa aphid in Lahontan alfalfa. J. Econ. Entomol. 50: 320–324.

Hoy, C.W., C. Jennison, A.M. Shelton, and J.T. Andaloro. 1983. Variable-intensity sampling—a new technique for decision-making in cabbage pest management. J. Econ. Entomol. 76: 139–143.

Hoy, M.A., ed. 1982. *Recent advances in knowledge of the Phytoseiidae*. Univ. Calif. Div. Agr. Sci. Publ. 3284. 92 pp.

Hoy, M.A. 1985. Recent advances in genetics and genetic improvements of the Phytoseiidae. Ann. Rev. Entomol. 30: 345–370.

Hoy, M.A., and D.C. Herzog, eds. 1985. *Biological control in agricultural IPM systems*. Academic Press, New York. 589 pp.

Hoy, M.A., and N.F. Knop. 1981. Selection for and genetic analysis of permethrin resistance in *Metaseiulus occidentalis*: genetic improvement of a biological control agent. Entomol. Exp. Appl. 30: 10–18.

Huffaker, C.B., ed. 1971. *Biological control*. Plenum, New York. 511 pp.

Huffaker, C.B., ed. 1980. *New technology of pest control*. Wiley, New York. 500 pp.

Huffaker, C.B., and P.S. Messenger. 1976. *Theory and practice of biological control*. Academic Press, New York. 788 pp.

Huffaker, C.B., F.J. Simmonds, and J.E. Liang. 1976. The theoretical and empirical basis of biological control. pp. 42–78 in C.B. Huffaker and P.S. Messenger, eds. *Theory and practice of biological control*. Academic Press, New York.

Hughes, R.D., M. Tyndale-Briscoe, and J. Walker. 1978. Effects of introduced dung beetles (Coleoptera: Scarabaeidae) on the breeding and abundance of the Australian bushfly *Musca vetustissima* Walker (Diptera: Muscidae). Bull. Entomol. Res. 63: 361–372.

Hull, L.A., and E.H. Beers. 1985. Ecological selectivity: modifying chemical control practices to preserve natural enemies. pp. 103–122 in M.A. Hoy and D.C. Herzog, eds. *Biological control in agricultural IPM systems*. Academic Press, New York.

Hussey, N.W., and L. Bravenboer. 1971. Control of insect pests in glasshouse culture by the introduction of natural enemies. pp. 195–216 in C.B. Huffaker, ed. *Biological control*. Plenum, New York.

Ignoffo, C.M. 1985. Manipulating enzootic–epizootic diseases of arthropods. pp. 243–262 in M.A. Hoy and D.C. Herzog, eds. *Biological control in agricultural IPM systems*. Academic Press, New York.

Ikeda, T., F. Matsumura, and D.M. Benjamin. 1977. Chemical basis for feeding adaptation of pine sawflies *Neodiprion rugifrons* and *N. swainei*. Science 197: 497–499.

Inouye, D.W. 1978. Resource partitioning in bumblebees: Experimental studies of foraging behavior. Ecology 59: 672–678.

Inouye, R.S. 1980. Stabilization of a predator–prey equilibrium by the addition of a second "keystone" victim. Amer. Nat. 115: 300–305.

Isely, D. 1942. *Methods of insect control*. Burgess, Minneapolis. 135 pp.

Ives, W.G.H. 1964. Problems encountered in the development of life tables for insects. Proc. Entomol. Soc. Manitoba 20: 34–44.

Iwao, S. 1968. A new regression method for analyzing the aggregation pattern of insect populations. Res. Popul. Ecol. 10: 1–20.

Iwao, S. 1975. A new method of sequential sampling to classify populations relative to a critical density. Res. Popul. Ecol. 16: 281–288.

Jackson, D.S., and B.G. Lee. 1985. Medfly in California 1980–1982. Bull. Entomol. Soc. Amer. 31(4): 29–37.

Jacobson, M. 1982. Plant, insects, and man—their interrelationships. Econ. Botany 36: 346–354.

Jansson, R.K., and Z. Smilowitz. 1986. Influence of nitrogen on population parameters of potato insects: Abundance, population growth, and within-plant distribution of the green peach aphid, *Myzus persicae* (Homoptera: Aphidae). Envir. Entomol. 15: 49–55.

Janzen, D.H. 1971. Escape of *Cassia grandis* L. beans from predators in time and space. Ecology 52: 964–979.

Jermy, T. 1984. Evolution of insect/host plant relationships. Amer. Nat. 124: 609–630.

Johnson, C.G. 1969. *Migration and dispersal of insects by flight*. Methuen, London. 763 pp.

Johnson, W.T., and H.H. Lyon. 1976. *Insects that feed on trees and shrubs*. Cornell University Press, Ithaca, NY. 464 pp.

Jones, C.G. 1983. Phytochemical variation, colonization, and insect communities: The case of bracken fern (*Pteridium aquilinum*). pp. 513–558 in R.F. Denno and M.S. McClure, eds. *Variable plants and herbivores in natural and managed systems*. Academic Press, New York.

Jones, V.P., and M.P. Parrella. 1984. The sublethal effects of selected insecticides on life table parameters of *Panonychus citri* (Acari: Tetranychidae). Canad. Entomol. 116: 1033–1040.

Jones, V.P., M.P. Parrella, and D.R. Hodel. 1986. Biological control of leafminers in greenhouse chrysanthemums. Calif. Agr. (Jan.-Feb.), pp. 10–12.

Joyce, R.J.V. 1959. The yield response of cotton in the Sudan Gezira to DDT spray. Bull. Entomol. Res. 50: 567–594.

Kalmakoff, J., and A.R. Miles. 1980. Ecological approaches to the use of microbial pathogens in insect control. BioScience 30: 344-347.

Karban, R. 1982. Increased reproductive success at high densities and predator satiation for periodical cicadas. Ecology 63: 321-328.

Karban, R. 1986. Induced resistance against spider mites in cotton: field verification. Entomol. Exp. Appl. 42: 239-242.

Karban, R., and J.R. Carey. 1984. Induced resistance of cotton seedlings to mites. Science 225: 53-54.

Kareiva, P. 1983a. Influence of vegetation texture on herbivore populations: Resource concentration and herbivore movement. pp. 259-289 in R.F. Denno and M.S. McClure, eds. *Variable plants and herbivores in natural and managed systems.* Academic Press, New York.

Kareiva, P. 1983b. Local movement in herbivorous insects: Applying a passive diffusion model to mark–recapture field experiments. Oecologia 57: 322-327.

Kareiva, P. 1985. Finding and losing host plants by *Phyllotreta*: Patch size and surrounding habitat. Ecology 66: 1809-1816.

Kaup, W.J., and S.S. Sohi. 1985. The role of viruses in the ecosystem. pp. 441-465. In K. Maramorosch and K.E. Sherman, eds. *Viral insecticides for biological control.* Academic Press, New York.

Kaya, H.K. 1985. Entomogenous nematodes for insect control in IPM systems. pp. 283-302 in M.A. Hoy and D.C. Herzog, eds. *Biological control in agricultural IPM systems.* Academic Press, New York.

Keh, B. 1983. Cryptic arthropod infestations and illusions and delusions of parasitoses. pp. 165-185 in G.W. Frankie and C.S. Koehler, eds. *Urban entomology: Interdisciplinary perspectives.* Praeger, New York.

Kempton, R.A., and L.R. Taylor. 1976. Models and statistics for species diversity. Nature 262: 818-820.

Kenaga, E.E., and C.S. End. 1978. *Commercial and experimental organic insecticides.* Entomol. Soc. Amer. Spec. Publ. 77 pp.

Kennedy, G.G. 1978. Recent advances in insect resistance of vegetable and fruit crops in North America: 1966-77. Bull. Entomol. Soc. Amer. 29: 375-384.

Kennedy, G.G., and D.C. Margolies. 1985. Mobile arthropod pests: Management in diversified agroecosystems. Bull. Entomol. Soc. Amer. 31(3): 21-27.

Kennedy, M.E. 1978. *Assessing the role of vertebrates in biological control of invertebrate populations.* Haywood Tech. Inst., Clyde, NC. 196 pp.

Khush, G.S. 1977. Breeding for resistance in rice. Ann. N. Y. Acad. Sci. 28: 296-308.

Kielbaso, J.J., and M.K. Kennedy. 1983. Urban forestry and entomology: A current appraisal. pp. 423-440 in G.W. Frankie and C.S. Koehler, eds. *Urban entomology: Interdisciplinary perspectives.* Praeger, New York.

Kimmerer, W.J. 1984. Diversity/stability: A criticism. Ecology 65: 1936-1938.

King, E.G., K.R. Hopper, and J.E. Powell. 1985. Analysis of systems for biological control of crop arthropod pests in the United States by augmentation of predators and parasites. pp. 201-227 in M.A. Hoy and D.C. Herzog, eds. *Biological control in agricultural IPM systems.* Academic Press, New York.

Klopfenstein, W.G. 1977. Effect of European red mite feeding on growth and yield of "red delicious" apple. PhD thesis, Ohio State University, Columbus.

Klun, J.A., and T.A. Brindley. 1966. Role of 6-methoxybenzoxazolinone in inbred resistance of host plant (maize) to first-brood larvae of European corn borer. J. Econ. Entomol. 59: 711-718.

Knipling, E.F. 1979. *The basic principles of insect population suppression and management.* USDA Agr. Handbook 512. 659 pp.

Knipling, E.F., and E.A. Stadelbacher. 1983. The rationale for areawide management of *Heliothis* (Lepidoptera: Noctuidae) populations. Bull. Entomol. Soc. Amer. 29(4): 29-37.

Kogan, M. 1982. Plant resistance in pest management. pp. 93-134 in R.L. Metcalf and W.L. Luckmann, eds. *Introduction to insect pest management*, 2nd ed. Wiley, New York.

Kogan, M., and E.F. Ortman. 1978. Antixenosis—a new term proposed to define Painter's "nonpreference" modality of resistance. Bull. Entomol. Soc. Amer. 24: 175-176.

Kogan, M., W.G. Ruesink, and K. McDowell. 1974. Spatial and temporal distribution patterns of the bean leaf beetle, *Cerotoma trifurcata* (Forster) on soybeans in Illinois. Envir. Entomol. 3: 607-617.

Komarek, E.V., and W.J. Kloft, eds. 1980. *The fire ant problem.* Proc. Tall Timbers Conf. Ecol. Anim. Contr. Hab. Man. 7. 173 pp.

Kouskolekas, C., and G.C. Decker. 1968. A quantitative evaluation of factors affecting alfalfa yield reduction caused by potato leafhopper attack. J. Econ. Entomol. 61: 921-927.

Krafsur, E.S. 1985. Screwworm flies (Diptera: Calliphoridae): Analysis of sterile mating frequencies and covariates. Bull. Entomol. Soc. Amer. 31 (3): 36-40.

Krischik, V.A., and R.F. Denno. 1983. Individual population, and geographic patterns in plant defense. pp. 463-512 in R.F. Denno and M.S. McClure, eds. *Variable plants and herbivores in natural and managed systems.* Academic Press, New York.

Krombein, K.V., P.D. Hurd, Jr., D.R. Smith, and B.D. Burks. 1979. *Catalog of Hymenoptera in America north of Mexico.* Smithsonian Inst. Press, Washington, DC. 3 vols.

Kulman, H.M. 1971. Effects of insect defoliation on growth and mortality of trees. Ann. Rev. Entomol. 16: 289-324.

Kumar, R. 1984. *Insect pest control with specific reference to African agriculture.* Edw. Arnold, London. 298 pp.

Lamb, R.J., and W.J. Turnock. 1982. Economics of insecticidal control of flea beetles (Coleoptera: Chrysomelidae) attacking rape in Canada. Canad. Entomol. 114: 827-840.

Laven, H. 1967. Eradication of *Culex pipiens fatigans* through cytoplasmic incompatibility. Nature 216: 383-384.

Law, J.H., and F.E. Regnier. 1971. Pheromones. Ann. Rev. Biochem. 40: 533-548.

Lawson, F.R., R.L. Rabb, F.E. Guthrie, and T.G. Bowery. 1961. Studies of an integrated control system for hornworms on tobacco. J. Econ. Entomol. 54: 93-97.

Lawton, J.H., and D.R. Strong, Jr. 1981. Community patterns and competition in folivorous insects. Amer. Nat. 118: 317-318.

Legner, E.F., and R.A. Medved. 1981. Pink bollworm, *Pectinophora gossypiella* (Diptera: Gelechiidae) suppression with gossyplure, a pyrethroid, and parasite releases. Canad. Entomol. 113: 355-357.

Leibee, G.L., and D.J. Horn. 1979. Effect of tillage on survivorship of parasitoids of the cereal leaf beetle. Envir. Entomol. 8: 485-486.

Leius, K. 1967. Influence of wild flowers on parasitism of tent caterpillar and codling moth. Canad. Entomol. 99: 444-446.

Lemmon, H. 1986. Comax: An expert system for cotton crop management. Science 233: 29-33.

Leonhardt, B.A., and M. Beroza, eds. 1982. *Insect pheromone technology: Chemistry and applications.* Amer. Chem. Soc., Washington, DC. 260 pp.

LeVeen, E.P., and W.R.Z. Willey. 1983. A political economic analysis of urban pest management. pp. 19–40 in G.W. Frankie and C.S. Koehler, eds. *Urban entomology: Interdisciplinary perspectives.* Praeger, New York.

Levenson, H., and G.W. Frankie. 1983. A study of homeowner attitudes and practices toward arthropod pests and pesticides in three U.S. metropolitan areas. pp. 67–106 in G.W. Frankie and C.S. Koehler, eds. *Urban entomology: Interdisciplinary perspectives.* Praeger, New York.

Levin, D.B., J.E. Laing, R.P. Jaques, and J.E. Corrigan. 1983. Transmission of the granulosis virus of *Pieris rapae* (Lepidoptera: Pieridae) by the parasitoid *Apanteles glomeratus* (Hymenoptera: Braconidae). Envir. Entomol. 12: 166–170.

Levine, E. 1977. Effect of the plum curculio, *Conotrachelus nenuphar* (Herbst), on apple and plum fruit abscission. PhD thesis, Ohio State University, Columbus.

Levinson, H.Z., and A.R. Levinson. 1985. Storage and insect species of stored grains and tombs in ancient Egypt. Z. Angew. Entomol. 100: 321–329.

Lewis, D.R. 1977. Life table analysis of alfalfa weevil population dynamics in Ohio 1973–1976. PhD thesis, Ohio State University, Columbus.

Lewis, T., and L.R. Taylor. 1967. *Introduction to experimental ecology.* Academic Press, New York. 401 pp.

Lewis, W.J. 1981. Semiochemicals: Their role in changing approaches to pest control. pp. 3–12 in D.A. Nordlund, R.L. Jones, and W.J. Lewis. *Semiochemicals: Their role in pest control.* Wiley, New York.

Lewontin, R. 1974. *The genetic basis of evolutionary change.* Columbia University Press, New York. 346 pp.

Lichtenberg, E.R., and W. Getz. 1985. Economics of rice-field mosquito control in California. BioScience 35: 292–297.

Lichtenstein, E.P., J.L. Kunstman, T.W. Fuhremann, and T.T. Liang. 1979. Effects of atrazine on the toxicity, penetration, and metabolism of carbofuran in the house fly. J. Econ. Entomol. 72: 785–789.

Lincoln, F.C. 1930. *Calculating waterfowl abundance on the basis of banding returns.* U.S. Dept. Agr. Circular 118. 4 pp.

Lloyd, M. 1967. "Mean crowding." J. Anim. Ecol. 36: 1–30.

Lotka, A.J. 1920. Analytical notes on certain rhythmic relations in organic systems. Proc. Nat. Acad. Sci. 7: 410–415.

Lowrance, R., B.R. Stinner, and G.J. House, eds. 1984. *Agricultural ecosystems: Unifying concepts.* Wiley, New York. 233 pp.

Luck, R.F. 1971. An appraisal of two methods of analyzing insect life tables. Canad. Entomol. 103: 1261–1271.

Luck, R.F., and D.E. Dahlsten. 1980. Within and between tree variation of live and parasitized Douglas-fir tussock moths, *Orgyria pseudotsugae* (Lepidoptera: Lymantriidae), cocoons on white fir in central California and its implications for sampling. Canad. Entomol. 112: 231–238.

Luck, R.F., and H. Podoler. 1985. Competitive exclusion of *Aphytis lignanensis* by *A. melinus*: Potential role of host size. Ecology 66: 904–913.

Luck, R.F., and G.T. Scriven. 1979. The elm leaf beetle, *Pyrrhalta luteola*, in southern California: Its host preference and host impact. Envir. Entomol. 8: 307–313.

Luck, R.F., R. van den Bosch, and R. Garcia. 1977. Chemical insect control—a troubled pest management strategy. BioScience 27: 606–611.

Luck, R.F., H. Podoler, and R. Kfir. 1982. Host selection and egg allocation behavior by *Aphytis melinus* and *A. lignanensis*: Comparison of two facultatively gregarious parasitoids. Ecol. Entomol. 7: 397–408.

Luna, J.M., S.J. Fleischer, and W.A. Allen. 1983. Development and validation of sequential sampling plans for potato leafhopper (Homoptera: Cicadellidae) in alfalfa. Envir. Entomol. 12: 1690–1694.

MacArthur, R.H., and E.O. Wilson. 1963. An equilibrium theory of insular biogeography. Evolution 17: 373–387.

MacArthur, R.H., and E.O. Wilson. 1967. *The theory of island biogeography*. Princeton University Press, Princeton, NJ. 203 pp.

Mack, T.P., B.A. Bajusz, E.S. Nolan, and Z. Smilowitz. 1981. Development of a temperature-mediated functional response equation. Envir. Entomol. 10: 573–579.

Maddox, J.V. 1982. Use of insect pathogens in pest management. pp. 175–216 in R.L. Metcalf and W.L. Luckmann, eds. *Introduction to insect pest management*, 2nd ed. Wiley, New York.

Madsen, H.F., and B.E. Carty. 1979. Codling moth (Lepidoptera: Olethreutidae) suppression by male removal with sex pheromone traps in three British Columbia orchards. Canad. Entomol. 111: 627–630.

Madsen, H.F., and B.J. Madsen. 1982. Populations of beneficial and pest arthropods in an organic and pesticide-treated apple orchard in British Columbia. Canad. Entomol. 114: 1083–1088.

Makela, M.E., and R.H. Richardson. 1978. Hidden, reproductively isolated populations: One of nature's countermeasures to genetic pest control. pp. 49–66 in R.H. Richardson, ed. *The screwworm problem*. University of Texas Press, Austin.

Malthus, T.R. 1798. *An essay on the principle of population*. Johnson, London. 505 pp.

Mangel, M., R.E. Plant, and J.R. Carey. 1984. Rapid delimiting of pest infestations: A case study of the Mediterranean fruit fly. J. Appl. Ecol. 31: 563–579.

Maramorosch, K., and K.F. Harris, eds. 1981. *Plant diseases and vectors: Ecology and epidemiology*. Academic Press, New York. 368 pp.

Maramorosch, K., and K.E. Sherman, eds. 1985. *Viral insecticides for biological control*. Academic Press, New York. 809 pp.

Marquis, R.J. 1984. Leaf herbivores decrease fitness of a tropical plant. Science 226: 537–539.

Marschall, K.J. 1970. Introduction of a new virus disease of the coconut rhinoceros beetle in western Samoa. Nature 225: 288–289.

Marshall, E. 1985. The rise and decline of Temik. Science 229: 1369–1371.

Martin, H., and D. Woodcock. 1983. *The scientific principles of crop protection*, 7th ed. Arnold, London. 486 pp.

Mason, R.R. 1976. Life tables for a declining population of the Douglas-fir tussock moth in northeastern Oregon. Ann. Entomol. Soc. Amer. 69: 948–958.

Mason, R.R. 1981. Numerical analysis of the causes of population collapse in a severe outbreak of the Douglas-fir tussock moth. Ann. Entomol. Soc. Amer. 74: 51–57.

Masud, S.M., R.D. Lacewell, C.R. Taylor, J.H. Beneaict, and L.A. Lippke. 1981. Economic impact of integrated pest management strategies for cotton production in the coastal bend region of Texas. Southern J. Agr. Econ. (Dec.), pp. 47–52.

Maxwell, F.G., and P.R. Jennings, eds. 1980. *Breeding plants resistant to insects*. Wiley, New York. 683 pp.

May, R.M. 1973. Time-delay versus stability in population models with two and three trophic levels. Ecology 54: 315–325.

May, R.M. 1975. *Stability and complexity in model ecosystems*. Princeton University Press, Princeton, NJ. 265 pp.

Mayo, O. 1980. *The theory of plant breeding*. Oxford University Press, New York. 293 pp.

McClure, M.S. 1980. Competition between exotic species: Scale insects in hemlock. Ecology 61: 1391–1401.

McClure, M.S. 1983. Competition between herbivores and increased resource heterogeneity. pp. 125-153 in R.F. Denno and M.S. McClure, eds. *Variable plants and herbivores in natural and managed systems.* Academic Press, New York.

McGaughey, W.H. 1985. Insect resistance to the biological insecticide *Bacillus thuringiensis.* Science 229: 193-195.

McGovern, T.P., M. Beroza, T.L. Ladd, Jr., J.C. Ingange, and J.P. Purimos. 1970. Phenethyl propionate, a potent new attractant for Japanese beetles. J. Econ. Entomol. 63: 1727-1729.

McGuckin, T. 1983. Alfalfa management strategies for Wisconsin dairy farms—application of stochastic dominance. North Centr. J. Agr. Econ. 5: 43-49.

McNiel, J. 1975. Juvenile hormone analogs: Detrimental effects on the development of an endoparasitoid. Science 189: 640-642.

Melander, A.L. 1914. Can insects become resistant to sprays? J. Econ. Entomol. 7: 167-173.

Mertins, J.W. 1986. Arthropods on the screen. Bull. Entomol. Soc. Amer. 32: 85-90.

Messenger, P.S. 1964. Use of life tables in a bioclimatic study of an experimental aphid-braconid wasp host-parasite system. Ecology 45: 119-131.

Messenger, P.S., E. Biliotti, and R. van den Bosch. 1976. The importance of natural enemies in integrated control. pp. 543-563 in C.B. Huffaker and P.S. Messenger, eds. *Theory and practice of biological control.* Academic Press, New York.

Metcalf, R.L. 1980. Changing role of insecticides in crop production. Ann. Rev. Entomol. 25: 219-256.

Metcalf, R.L. 1982. Insecticides in pest management. pp. 217-277 in R.L. Metcalf and W.L. Luckmann, eds. *Introduction to insect pest management,* 2nd ed. Wiley, New York.

Metcalf, R.L., and W.L. Luckmann, eds. 1982. *Introduction to insect pest management,* 2nd ed. Wiley, New York. 577 pp.

Metcalf, R.L., W.P. Flint, and C.L. Metcalf. 1962. *Destructive and useful insects,* 4th ed. McGraw-Hill, New York. 1087 pp.

Michalson, E.L. 1975. Economic impact of mountain pine beetle on outdoor recreation. So. J. Agr. Econ. (Dec.), pp. 43-50.

Michelbacher, A.E. 1940. Effect of *Bathyplectes curculionis* on the alfalfa weevil population in lowland middle California. Hilgardia 13: 81-99.

Miller, J.C. 1983. Ecological relationships among parasites and the practice of biological control. Envir. Entomol. 12: 620-624.

Miller, J.R., and T.A. Miller, eds. 1986. *Insect-plant interactions.* Springer-Verlag, New York. 342 pp.

Miller, T.E. 1982. Community diversity and interactions between the size and frequency of disturbance. Amer. Nat. 120: 533-536.

Milne, A. 1962. On the theory of natural control of insect populations. J. Theoret. Biol. 3: 19-50.

Mitchell, H.C., W.H. Cross, W.L. McGovern, and E.M. Dawson. 1973. Behavior of the boll weevil on frego bract cotton. J. Econ. Entomol. 66: 677-680.

Moffit, H.R., K.D. Mantey, and G. Tamaki. 1984. Effects of residues of chitin-synthesis inhibitors on egg hatch and subsequent larval entry of the codling moth *Cydia pomonella* (Lepidoptera: Olethreutidae). Canad. Entomol. 116: 1057-1062.

Moon, R.D. 1980. Biological control through interspecific competition. Envir. Entomol. 9: 723-728.

Mooney, H.A., and S.L. Gulmon. 1982. Constraints on leaf structure and function in reference to herbivory. BioScience 32: 198-206.

Morris, O.N. 1977. Long-term effects of aerial applications of virus–fenitrothion combinations against the spruce budworm, *Choristoneura fumiferana* (Lepidoptera: Tortricidae). Canad. Entomol. 109: 9-14.

Morris, O.N. 1980. Entomopathogenic viruses: Strategies for use in forest integrated pest management. Canad. Entomol. 112: 573-584.

Morris, O.N. 1985. Susceptibility of the migratory grasshopper, *Melanoplus sanguinipes* (Orthoptera: Acrididae) to mixtures of *Nosema locustae* (Microsporidia: Nosematidae) and chemical insecticides. Canad. Entomol. 117: 131-132.

Morris, R.F. 1959. Single-factor analysis in population dynamics. Ecology 45: 119-131.

Morris, R.F. 1960. Sampling insect populations. Ann. Rev. Entomol. 5: 243-264.

Morris, R.F., ed. 1963. The dynamics of epidemic spruce budworm populations. Mem. Entomol. Soc. Canada #31, 332 pp.

Morris, R.F. 1972. Predation by wasps, birds and mammals on *Hyphantrea cunea*. Canad. Entomol. 104: 1581-1591.

Morrow, P.A., T.E. Bellas, and T. Eisner. 1976. Eucalyptus oils in the defensive oral discharge of Australian sawfly larvae (Hymenoptera: Pergidae). Oecologia 24: 193-206.

Mound, L.A. 1965. Effect of leaf hair on cotton whitefly population in the Sudan Gezira. Emp. Cotton Grow. Rev. 42: 33-40.

Muldrew, J.A. 1953. The natural immunity of the larch sawfly (*Pristophora erichsonii* (Htg.)) to the introduced parasite *Mesoleius tenthredinis* Morley, in Manitoba and Saskatchewan. Can. J. Zool. 31: 313-332.

Mullin, C.A., and B.A. Croft. 1985. An update on development of selective pesticides favoring arthropod natural enemies. pp. 123-150 in M.A. Hoy and D.C. Herzog, eds. *Biological control in agricultural IPM systems*. Academic Press, New York.

Mullin, C.A., B.A. Croft, K. Strickler, F. Matsumura, and J.R. Miller. 1982. Detoxification enzyme differences between a herbivorous and predatory mite. Science 217: 1270-1272.

Mumford, J.D. 1981. A study of sugar beet growers' pest control decisions. Ann. Appl. Biol. 97: 243-252.

Mumford, J.D., and G.A. Norton. 1984. Economics of decision making in pest management. Ann. Rev. Entomol. 29: 157-174.

Murdoch, W.W. 1975. Diversity, complexity, stability and pest control. J. Appl. Ecol. 12: 795-807.

Murdoch, W.W., and J.R. Marks. 1973. Predation by coccinellid beetles: Experiments on switching. Ecology 54: 160-167.

Murdoch, W.W., J.D. Reeve, C.B. Huffaker, and C.E. Kennett. 1984. Biological control of olive scale and its relevance to ecological theory. Amer. Nat. 123: 371-392.

Murdoch, W.W., J. Chesson, and P.L. Chesson. 1985. Biological control in theory and practice. Amer. Nat. 125: 344-366.

Nasci, R.S., C.W. Harris, and C.K. Porter. 1983. Failure of an insect electrocuting device to reduce mosquito biting. Mosq. News 43: 180-184.

Navarajan, P.A.V., B. Parshad, R. Ahmed, and R. Dass. 1979. Effect of some insecticides on parasitism by the parasitoid *Trichogramma brasiliensis* (Ashmead). Z. Angew. Entomol. 88: 399-403.

Neuvonen, S., and E. Haukioja. 1985. How to study induced plant resistance? Oecologia 66: 456-457.

Nevo, E. 1978. Genetic variation in natural populations: Pattern and theory. Theor. Pop. Biol. 13: 121-177.

Newsom, L.D., M. Kogan, F.D. Miner, R.L. Rabb, S.G. Turnipseed, and W.H. Whitcomb. 1980. General accomplishments toward better pest control in soybeans.

pp. 51–98 in C.B. Huffaker, ed. *New technology of pest control.* Wiley, New York.

Nielson, M.M. 1965. Effects of cytoplasmic polyhedrosis on adult Lepidoptera. J. Invert. Pathol. 7: 306–314.

Niles, G.A. 1980. Breeding cotton for resistance to insect pests. pp. 337–369 in F.G. Maxwell and P.R. Jennings, eds. *Breeding plants resistant to insects.* Wiley, New York.

Nishida, R., W.S. Bowers, and P.H. Evans. 1983. Juvadecene: Discovery of a juvenile hormone mimic in the plant *Macropiper excelsum.* Arch. Ins. Bioch. Physiol. 1: 17–24.

Nordin, G.L., G.C. Brown, and J.A. Millstein. 1983. Epizootic phenology of *Erynia* disease of the alfalfa weevil, *Hypera postica,* in central Kentucky. Envir. Entomol. 12: 1350–1355.

Nordlund, D.A., and W.J. Lewis. 1976. Terminology of chemical releasing stimuli in intraspecific and interspecific interactions. J. Chem. Ecol. 2: 211–220.

Nordlund, D.A., R.L. Jones, and W.J. Lewis. 1981. *Semiochemicals: Their role in pest control.* Wiley, New York, 306 pp.

Nordlund, D.A., R.B. Chalfant, and W.J. Lewis. 1985. Behavior-modifying chemicals to enhance natural enemy effectiveness. pp. 89–101 in M.A. Hoy and D.C. Herzog, eds. *Biological control in agricultural IPM systems.* Academic Press, New York.

Norgaard, R.B. 1976. The economics of improving pesticide use. Ann. Rev. Entomol. 21: 45–60.

Norris, D.M., and M. Kogan. 1980. Biochemical and morphological bases of resistance. pp. 23–61 in F.G. Maxwell and P.R. Jennings, eds. *Breeding plants resistant to insects.* Wiley, New York.

Norton, G.A., and G.R. Conway. 1977. The economic and social context of pest, disease, and weed problems. pp. 205–226 in J.M. Cherrett and G.R. Sagar, eds. *Origin of pest, parasite, disease and weed problems.* Blackwell, London.

Nyrop, J.P., and G.A. Simmons. 1984. Errors incurred when using Iwao's sequential decision rule in insect sampling. Envir. Entomol. 13: 1459–1465.

Oka, I.N., and D. Pimentel. 1974. Corn susceptibility to corn leaf aphids and common corn smut after herbicide treatment. Envir. Entomol. 3: 911–915.

Olfert, O.O., M.K. Mukerji, and J.F. Doane. 1985. Relationship between infestation levels and yield loss caused by wheat midge, *Sitodiplosis moselana* (Gehin) (Diptera: Cecidomyiidae) in spring wheat in Saskatchewan. Canad. Entomol. 117: 593–598.

Olkowski, H., and W. Olkowski. 1976. Entomophobia in the urban ecosystem, some observations and suggestions. Bull. Entomol. Soc. Amer. 22: 313–317.

Olkowski, W. 1974. A model ecosystem management program. Proc. Tall Timbers Conf. Ecol. Anim. Contr. Hab. Man. 5: 103–117.

Onsager, J.A. 1976. *The rationale of sequential sampling with emphasis on its use in pest management.* U.S. Dept. Agr. Tech. Bull. 1526. 16 pp.

Onstad, D.W., C.A. Shoemaker, and B.C. Hansen. 1984. Management of potato leafhopper, *Empoasca fabae* (Homoptera: Cicadellidae) on alfalfa with the aid of systems analysis. Envir. Entomol. 13: 1046–1058.

Ortman, E.E., and D.C. Peters. 1980. Introduction, pp. 3–21 in F.G. Maxwell and P.R. Jennings, eds. *Breeding plants resistant to insects.* Wiley, New York.

Otvos, I.S., and R.W. Stark. 1985. Arthropod food of some forest-inhabiting birds. Canad. Entomol. 117: 971–990.

Paine, T.D., M.C. Birch, and P. Svirha. 1981. Niche breadth and resource partitioning by four sympatric species of bark beetles. Oecologia 48: 1–6.

Painter, R.H. 1951. *Insect resistance in crop plants.* Macmillan, New York. 520 pp.

Parrella, M.P., and C.B. Keil. 1984. Insect pest management: The lesson of *Liromyza.* Bull. Entomol. Soc. Amer. 30 (2): 22–25.

Parry, G.D. 1981. The meanings of *r*- and *K*-selection. Oecologia 48: 260–264.

Pashley, D.P., and G.L. Bush. 1979. The use of allozymes in studying insect management with special reference to the codling moth *Laspeyresia pomonella* (Lepidoptera: Olethreutidae). pp. 333–341 in R.L. Rabb and G.G. Kennedy, eds. *Movement of highly mobile insects.* North Carolina State University, Raleigh.

Pashley, D.P., S.J. Johnson, and A.N. Sparks. 1985. Genetic population structure of migratory moths: The fall armyworm (Lepidoptera: Noctuidae). Ann. Entomol. Soc. Amer. 78: 756–762.

Patterson, R.S., D.E. Weidhaas, H.R. Ford, and C.S. Lofgren. 1970. Suppression and elimination of an island population of *Culex pipiens quinquefasciatus* with sterile males. Science 168: 1368–1370.

Patterson, R.S., G.C. LeBrecque, D.F. Williams, and D.E. Weidhaas. 1981. Control of the stable fly, *Stomoxys calcitrans* (Diptera: Muscidae) on St. Croix, U.S. Virgin Islands, using IPM measures. J. Med. Entomol. 18: 203–210.

Peacock, J.W., R.A. Cuthbert, and G.W. Lanier. 1981. Deployment of traps in a barrier strategy to reduce populations of the European elm bark beetle, and the incidence of Dutch elm disease. pp. 15–174 in E.R. Mitchell, ed. *Management of insect pests with semiochemicals.* Plenum, New York.

Pedigo, L.P., and J.W. van Schaik. 1984. Time-sequential sampling: A new use of the sequential probability ratio test for pest management decisions. Bull. Entomol. Soc. Amer. 30 (1): 32–36.

Pedigo, L.P., G.L. Lentz, J.D. Stone, and D.F. Cox. 1972. Green cloverworm populations in Iowa soybeans with special reference to sampling procedures. J. Econ. Entomol. 65: 414–421.

Pedigo, L.P., S.H. Hutchins, and L.G. Higley. 1986. Economic injury levels in theory and practice. Ann. Rev. Entomol. 31: 341–368.

Perkins, J.H. 1982. *Insects, experts, and the insecticide crisis. The quest for new management strategies.* Plenum, New York. 304 pp.

Person, C.O., D.J. Samborski, and R. Rohringer. 1952. The gene for gene concept. Nature 194: 561–562.

Petersen, J.J. 1982. Current status of nematodes for the biological control of insects. Parasitol. 84: 177–204.

Pfadt, R.E., ed. 1985. *Fundamentals of applied entomology,* 4th ed. Macmillan, New York. 798 pp.

Phillips, J.R., A.P. Gutierrez, and P.L. Adkisson. 1980. General accomplishments toward better insect control in cotton. pp. 123–153 in C.B. Huffaker, ed. *New technology of pest control.* Wiley, New York.

Piedrahita, O., C.R. Ellis, and O.B. Allen. 1985. Effect of spacing and clumping of corn plants on density of corn-rootworm larvae. Canad. Entomol. 117: 139–142.

Pielou, E.C. 1977. *Mathematical ecology.* Wiley, New York. 385 pp.

Pieters, E.P. 1978. Bibliography of sequential sampling plans for insects. Bull. Entomol. Soc. Amer. 24: 372–374.

Pillemer, E.A., and W.M. Tingey. 1976. Hooked trichomes: A physical plant barrier to a major agricultural pest. Science 193: 482–484.

Pimentel, D. 1961a. On a genetic feed-back mechanism regulating populations of herbivores, parasites and predators. Amer. Nat. 95: 65–79.

Pimentel, D. 1961b. An ecological approach to the insecticide problem. J. Econ. Entomol. 54: 108–114.

Pimentel, D. 1963. Introducing parasites and predators to control native pests. Canad. Entomol. 95: 785-792.

Pimentel, D. 1973. Extent of pesticide use, food supply, and pollution. J. New York Entomol. Soc. 81: 13-37.

Pimentel, D., ed. 1981. *CRC handbook of pest management in agriculture*, vol. I. CRC Press, Boca Raton, FL. 600 pp.

Pimentel, D., and M. Burgess. 1985. Effects of single versus combinations of insecticides on the development of resistance. Envir. Entomol. 14: 582-589.

Pimentel, D. and L. Levitan. 1986. Pesticides: Amounts applied and amounts reaching pests. BioScience 36: 86-91.

Pimentel, D., E.C. Terhune, W. Dritschilo, D. Gallahan, N. Kinner, D. Nafus, R. Peterson, N. Zareh, J. Mistiti, and O. Haber-Schaim. 1977. Pesticides, insects in foods, and cosmetic standards. BioScience 27: 178-185.

Pimentel, D., J. Krummel, D. Gallahan, J. Hough, A. Merrill, I. Schreiner, P. Vittum, F. Koziol, E. Back, D. Yen, and S. Fiance. 1978. Benefits and risks of pesticide use. BioScience 28: 772, 778-784.

Pimentel, D., D. Andow, R. Dyson-Hudson, D. Gallahan, S. Jacobson, M. Irish, S. Kroop, A. Moss, I. Schreiner, M. Shepard, T. Thompson, and B. Vinzant. 1980. Environmental and social costs of pesticides: A preliminary assessment. Oikos 34: 126-140.

Pimentel, D., C. Glenister, S. Fast, and D. Gallahan. 1984. Environmental risks of biological pest controls. Oikos 42: 283-290.

Plant, R.E. 1986. Uncertainty and the economic threshold. J. Econ. Entomol. 79: 1-6.

Plimmer, J.R., B.A. Leonhardt, R.E. Webb, and C.P. Schwalbe. 1982. Management of the gypsy moth with its sex attractant pheromone. pp. 231-242 in B.A. Leonhardt and M. Beroza, eds. *Insect pheromone technology: Chemistry and applications.* Amer. Chem. Soc., Washington, DC.

Plowright, R.C., and F.H. Rodd. 1980. The effect of aerial insecticide spraying on hymenopterous pollinators in New Brunswick. Canad. Entomol. 112: 259-269.

Podgwaite, J.D. 1985. Strategies for field use of baculoviruses. pp. 775-797 in K. Maramorosch and K.E. Sherman, eds. *Viral insecticides for biological control.* Academic Press, New York.

Podoler, H., and D. Rogers. 1975. A new method for the identification of key factors from life-table data. J. Anim. Ecol. 45: 85-114.

Podoler, H., and Z. Mendel. 1979. Analysis of a host–parasite (*Ceratitis-Muscidifurax*) relationship under laboratory conditions. Ecol. Entomol. 4: 45-59.

Poinar, G.O., Jr. 1972. Nematodes as facultative parasites of insects. Ann. Rev. Entomol. 17: 103-122.

Poinar, G.O., Jr. 1979. *Nematodes for biological control of insects.* CRC Press, Boca Raton, FL. 277 pp.

Poinar, G.O., Jr., and G.M. Thomas. 1984. *Laboratory guide to insect pathogens and parasites.* Plenum, New York. 392 pp.

Polis, G.A. 1981. The evolution and dynamics of intraspecific predation. Ann. Rev. Ecol. Syst. 12: 225-251.

Poston, F.L., L.P. Pedigo, and S.M. Welch. 1983. Economic injury levels: Reality and practicality. Bull. Entomol. Soc. Amer. 29 (1): 49-53.

Powell, W., G.J. Dean, and R. Bardner. 1985. Effects of pirimicarb, dimethoate, and benomyl on natural enemies of cereal aphids in winter wheat. Ann. Appl. Biol. 106: 235-242.

Prestwich, G.D., A.K. Gayen, S. Phirwa, and T.B. Kline. 1983. 29-fluorophytosteroids: Novel pro-insecticides which cause death by dealkylation. Biotechnology 1: 62-65.

Price, P.W. 1970. Trail odors: Recognition by insects parasitic on cocoons. Science 170: 546–547.

Price, P.W. 1976. Colonization of crops by arthropods: Nonequilibrium communities in soybeans. Envir. Entomol. 5: 605–611.

Price, P.W. 1983. Hypotheses on organization and evolution in herbivorous insect communities. pp. 559–596 in R.F. Denno and M.S. McClure, eds. *Variable plants and herbivores in natural and managed systems.* Academic Press, New York.

Price, P.W. 1984. *Insect ecology,* 2nd ed. Wiley, New York. 607 pp.

Price, P.W., C.N. Slobodchikoff, and W.S. Gaud. 1984. *A new ecology: Novel approaches to interactive systems.* Wiley, New York. 515 pp.

Proverbs, M.D., J.R. Newton, and C.J. Campbell. 1982. Codling moth: A pilot program of control by sterile insect release in British Columbia. Canad. Entomol. 114: 363–376.

Pruess, K.P. 1983. Day-degree methods for pest management. Envir. Entomol. 12: 613–619.

Rabb, R.L., and F.E. Guthrie, eds. 1970. *Concepts of pest management.* North Carolina State University Press, Raleigh. 242 pp.

Rabb, R.L., R.E. Stinner, and R. van den Bosch. 1976. Conservation and augmentation of natural enemies. pp. 233–254 in C.B. Huffaker and P.S. Messenger, eds. *Theory and practice of biological control.* Academic Press, New York.

Rainey, R.C. 1982. Putting insects on the map: Spatial inhomogeneity and the dynamics of insect populations. Antenna 6: 162–169.

Ramoska, W.A., G.R. Stairs, and W.F. Hink. 1975. Ultraviolet activation of insect nuclear polyhedrosis virus. Nature 253: 628–629.

Rathcke, B.J. 1976. Competition and coexistence within a guild of herbivorous insects. Ecology 57: 76–87.

Raupp, M.J. 1985. Effects of leaf toughness on mandibular wear of the leaf beetle *Plagoiodera versicolora.* Ecol. Entomol. 10: 73–79.

Raupp, M.J., and R.F. Denno. 1983. Leaf age as a predictor of herbivore distribution and abundance. pp. 91–124 in R.F. Denno and M.S. McClure, eds. *Variable plants and herbivores in natural and managed systems.* Academic Press, New York.

Rausher, M.D. 1983. Ecology of host-selection behavior in phytophagous insects. pp. 223–257 in R.F. Denno and M.S. McClure, eds. *Variable plants and herbivores in natural and managed systems.* Academic Press, New York.

Read, D.P., P.P. Feeny, and R.B. Root. 1970. Habitat selection by the aphid parasite *Diaeretiella rapae* (Hymenoptera: Braconidae) and hyperparasite *Charips brassicae* (Hymenoptera: Cynipidae). Canad. Entomol. 102: 1567–1578.

Reader, P.M., and T.R.E. Southwood. 1981. The relationship between palatability to invertebrates and the successional status of a plant. Oecologia 51: 271–275.

Rees, C.J.C. 1969. Chemoreceptor specificity associated with choice of feeding site by the beetle, *Chrysolina brunsvicensis* on its food plant, *Hypericum hirsutum.* pp. 565–583 in J. de Wilde and L.M. Schoonhoven, eds. *Insect and host plant.* North Holland, Amsterdam.

Reese, J.C. 1979. Interaction of allelochemicals with nutrients in herbivore food. pp. 309–330 in G.A. Rosenthal and D.H. Janzen, eds. *Herbivores, their interaction with secondary plant metabolites.* Academic Press, New York.

Reynolds, H.T., P.L. Adkisson, R.F. Smith, and R.E. Frisbie. 1982. Cotton insect pest management. pp. 375–441 in R.L. Metcalf and W.L. Luckmann, eds. *Introduction to insect pest management.* Wiley, New York.

Rhoades, D.F. 1983. Herbivore population dynamics and plant chemistry. pp. 155–220 in R.F. Denno and M.S. McClure, eds. *Variable plants and herbivores in natural and managed systems.* Academic Press, New York.

Richards, O.W. 1961. The theoretical and practical study of natural insect populations. Ann. Rev. Entomol. 6: 147–162.

Richardson, R.H., J.R. Ellison, and W.W. Averhoff. 1982. Autocidal control of screwworms on North America. Science 215: 361–370.

Riechert, S.E., and T. Lockley. 1984. Spiders as biological control agents. Ann. Rev. Entomol. 29: 299–320.

Riedl, H., and B.A. Croft. 1974. A study of pheromone trap catches in relation to codling moth damage. Canad. Entomol. 106: 525–537.

Ripper, W.E., and M. George. 1965. *Cotton pests of the Sudan.* Blackwell, Oxford. 345 pp.

Risch, S. 1981. Insect herbivore abundance in tropical monocultures and polycultures: An experimental test of two hypotheses. Ecology 62: 1325–1340.

Risebrough, R.W. 1986. Pesticides and bird populations. pp. 397–427 in R.F. Johnston, ed. *Current ornithology*, vol. 3. Plenum, New York.

Risser, P.G. 1985. Toward a holistic management perspective. BioScience 35: 414–418.

Roelofs, W.L. 1981. Attractive and aggregating pheromones. pp. 215–235 in D.A. Nordlund, R.L. Jones, and W.J. Lewis, eds. 1981. *Semiochemicals: Their role in pest control.* Wiley, New York. 306 pp.

Rogoff, W.M. 1980. Behavior modification for insect management by competitive displacement. Bull. Entomol. Soc. Amer. 26: 121–125.

Rohwer, G.C., and D.L. Williamson. 1983. Pest risk evaluation in regulatory entomology. Bull. Entomol. Soc. Amer. 29 (4): 41–46.

Rojas, B.A. 1964. La binomial negative y la estimación de intensidad de plagas en el suelo. Fitotech. Latinamerica 1: 27–36.

Root, R.B. 1973. Organization of a plant-arthropod association in simple and diverse habitats: The fauna of collards. Ecol. Monogr. 43: 95–124.

Root, R.B. 1975. Some consequences of ecosystem texture. pp. 83–97 in S.A. Levin, ed. *Ecosystem analysis and prediction.* Soc. Indus. Appl. Math., Philadelphia.

Root, R.B., and J.J. Skelsey. 1969. Biotic factors involved in crucifer aphid outbreaks. J. Econ. Entomol. 62: 223–233.

Rosenthal, G.A., and D.H. Janzen. 1979. *Herbivores: Their interaction with secondary plant metabolites.* Academic Press, New York. 718 pp.

Ross, M.H., K.B. Keil, and D.G. Cochran. 1981. The release of sterile males into natural populations of the German cockroach, *Blattella germanica.* Entomol. Exp. Appl. 30: 246–253.

Roush, R.T., and M.A. Hoy. 1980. Sevin resistance in spider mite predator. Calif. Agr. 34: 11–14.

Royama, T. 1981a. Fundamental concepts and methodology for the analysis of animal population dynamics with particular reference to univoltine species. Ecol. Monogr. 51: 473–493.

Royama, T. 1981b. Evaluation of mortality factors in insect life table analysis. Ecol. Monogr. 51: 495–505.

Ruesink, W.G. 1976. Status of the systems approach to pest management. Ann. Rev. Entomol. 21: 27–44.

Ruesink, W.G. 1980. Introduction to sampling theory. pp. 61–78 in M. Kogan and D.C. Herzog, eds. *Sampling methods in soybean entomology.* Springer-Verlag, New York.

Ruesink, W.G. 1982. Analysis and modelling in integrated pest management. pp. 353–373 in R.L. Metcalf and W.F. Luckmann, eds. *Introduction to insect pest management,* 2nd ed. Wiley, New York.

Ruesink, W.G., C.A. Shoemaker, A.P. Gutierrez, and G.W. Fick. 1980. The systems approach to research and decision making for alfalfa pest control. pp. 217–247 in C.B. Huffaker, ed. *New technology of pest control*. Wiley, New York.

Rust, M.K., and D.A. Reierson. 1983. Bibliography of ultrasound production and perception in insects. Bibliogr. Entomol. Soc. Amer. 2: 57–66.

Ryan, C.A. 1983. Insect-induced chemical signals regulating natural plant protection responses. pp. 43–60 in R.F. Denno and M.S. McClure, eds. *Variable plants and herbivores in natural and managed systems*. Academic Press, New York.

Ryan, R.B. 1983. Population density and dynamics of larch casebearer (Lepidoptera: Coleophoridae) in the Blue Mountains of Oregon and Washington before the build-up of exotic parasites. Canad. Entomol. 115: 1095–1102.

Salama, H.S., M.S. Foda, and A. Sharaby. 1985. Potential of some chemicals to increase the effectiveness of *Bacillus thuringiensis* Berl. against *Spodoptera littoralis* (Boisd.). Z. Angew. Entomol. 100: 425–433.

Sances, F.V., N.C. Toscano, M.W. Johnson, and L.F. LaPre. 1981. Pesticides may reduce lettuce yields. Calif. Agr. 11–12, pp. 4–5.

Santos, M.A. 1976. Prey selectivity and switching response of *Zetziella mali*. Ecology 57: 390–394.

Sauls, C.E., D.A. Nordlund, and W.J. Lewis. 1979. Kairomones and their use for management of entomophagous insects. VIII. Effect of diet on the kairomonal activity of frass from *Heliothis zea* (Boddie) larvae for *Microplitis croceipes* (Cresson). J. Chem. Ecol. 5: 363–369.

Sawyer, A.J., and R.A. Casagrande. 1983. Urban pest management: A conceptual framework. Urban Ecol. 7: 145–157.

Schmutterer, H. 1985. Which pests can be controlled by applications of neem seed kernel extracts under field conditions? Z. Angew. Entomol. 100: 468–475.

Schoener, T.W. 1982. The controversy over interspecific competition. Amer. Sci. 70: 586–595.

Schreiber, R.W. 1980. The brown pelican: An endangered species? BioScience 30: 42–47.

Schultz, J.C. 1983. Habitat selection and foraging tactics of caterpillars in heterogeneous trees. pp. 61–90 in R.F. Denno and M.S. McClure, eds. *Variable plants and herbivores in natural and managed systems*. Academic Press, New York.

Scriber, J.M. 1983. Evolution of feeding specialization, physiological efficiency, and host races in selected Papilionidae and Saturniidae. pp. 373–412 in R.F. Denno and M.S. McClure, eds. *Variable plants and herbivores in natural and managed systems*. Academic Press, New York.

Scriber, J.M. 1984. Host-plant suitability. pp. 159–202 in W.J. Bell and R.T. Carde, eds. *Chemical ecology of insects*. Chapman and Hall, London.

Sears, M.K., R.P. Jaques, and J.E. Laing. 1983. Utilization of action thresholds for microbial and chemical control of lepidopterous pests (Lepidoptera: Noctuidae, Pieridae) on cabbage. J. Econ. Entomol. 76: 368–374.

Seawright, J.A., P.E. Kaiser, D.A. Dame, and C.S. Lofgren. 1978. Genetic method for the preferential elimination of females of *Anopheles albimanus*. Science 200: 1303–1304.

Shepherd, R.F. 1985. Pest management of Douglas-fir tussock moth: Estimating larval density by sequential sampling. Canad. Entomol. 117: 1111–1115.

Shepherd, R.F., I. Otvos, and R.J. Chorney. 1984. Pest management of Douglas-fir tussock moth: a sequential sampling method to determine egg mass density. Canad. Entomol. 116: 1041–1049.

Shepherd, R.F., T.G. Gray, R.J. Chorney, and E.G. Daterman. 1985. Pest management of Douglas-fir tussock moth, *Orgyia pseudotsugata* (Lepidoptera: Lymantriidae): Monitoring endemic populations with pheromone traps to detect incipient outbreaks. Canad. Entomol. 117: 839–848.

Sherman, K.E. 1985. Considerations in the large-scale and commercial production of viral insecticides. pp. 757–774 in K. Maramorosch and K.E. Sherman, eds. *Viral insecticides for biological control*. Academic Press, New York.

Shoemaker, C. 1973. Optimization of agricultural pest management I: Biological and mathematical background. Math. Biosci. 16: 143–175.

Shoemaker, C. 1980. The role of systems analysis in integrated pest managment. pp. 25–49 in C.B. Huffaker, ed. *New technology of pest control*. Wiley, New York.

Shoemaker, C.A., and D.W. Onstad. 1983. Optimization analysis of the integration of biological, cultural, and chemical control of the alfalfa weevil (Coleoptera: Curculionidae). Envir. Entomol. 12: 286–295.

Silverstein, R.M. 1981. Pheromones: Background and potential for use in insect pest control. Science 213: 1326–1332.

Simberloff, D. 1976. Species turnover and equilibrium island biogeography. Science 194: 572–578.

Simberloff, D.L., and E.O. Wilson. 1969. Experimental zoogeography of islands. The colonization of empty islands. Ecology 50: 278–296.

Skovmand, O., and H. Mourier. 1986. Electrocuting light traps evaluated for the control of house flies. Z. Angew. Entomol. 102: 446–455.

Sláma, K. 1979. Insect hormones and antihormones in plants. pp. 683–700 in G.A. Rosenthal and D.H. Janzen, eds. *Herbivores: Their interaction with secondary plant metabolites*. Academic Press, New York.

Sláma, K., and C.M. Williams. 1965. Juvenile hormone activity for the bug *Pyrrhocoris apterus*. Proc. Nat. Acad. Sci. USA 54: 411–414.

Smiley, J.T., J.M. Horn, and N.E. Rank. 1985. Ecological effects of salicilin at three trophic levels: New problems from old adaptations. Science 229: 649–651.

Smilowitz, Z. 1981. GPA-CAST: A computerized model for green peach aphid management on potatoes. pp. 193–203 in J.H. Lashomb and R.A. Casagrande, eds. *Advances in potato pest management*. Academic Press, New York.

Smirnoff, W.A. 1972. Sensibilité de *Archips cerasivoranus* (Lepidoptera: Tortricidae) à l'infection par *Bacillus thuringiensis*. Canad. Entomol. 104: 1153–1159.

Smith, R.F. 1969. The new and the old in pest control. Proc. Accad. Nazion. Lincei, Roma 366: 21–30.

Smith, R.F., P.L. Adkisson, C.B. Huffaker, and L.D. Newsom. 1974. Progress achieved in the implementation of integrated control projects in the USA and tropical countries. EPPO Bull. 4: 221–239.

Southwood, T.R.E. 1961. The number of species of insects associated with various trees. J. Anim. Ecol. 30: 1–8.

Southwood, T.R.E. 1975. The dynamics of insect populations. pp. 151–199 in D. Pimentel, ed. *Insects, science, and society*. Academic Press, New York.

Southwood, T.R.E. 1977. Entomology and mankind. Amer. Sci. 65: 30–39.

Southwood, T.R.E. 1978. *Ecological methods with particular reference to the study of insect populations*, 2nd ed. Chapman and Hall, London. 524 pp.

Southwood, T.R.E., and H.N. Comins. 1976. A synoptic population model. J. Anim. Ecol. 45: 949–965.

Sower, L.L., G.E. Daterman, W. Funkhouser, and C. Sartwell. 1983. Pheromone disruption controls Douglas-fir tussock moth (Lepidoptera: Lymantriidae) reproduction at high insect densities. Canad. Entomol. 115: 965–969.

Staal, G.B. 1986. Anti-juvenile hormone agents. Ann. Rev. Entomol. 31: 391–429.

Stanton, M.L. 1983. Spatial patterns in the plant community and their effects upon insect search. pp. 125-157 in S. Ahmad, ed. *Herbivorous insects: Host-seeking behavior and mechanisms.* Academic Press, New York.

Stark, R.W. 1965. Recent trends in forest entomology. Ann. Rev. Entomol. 10: 303-324.

Stark, R.W. *et al.* 1980. Approach to research and forest management for mountain pine beetle control. pp. 397-416 in C.B. Huffaker, ed. *New technology of pest control.* Wiley, New York.

Steiner, L.F., W.C. Mitchell, E.J. Harris, T.T. Kozuma, and M.S. Fujimoto. 1965. Oriental fruit fly eradication by male annihilation. J. Econ. Entomol. 58: 961-964.

Stenseth, N.C. 1981. How to control pest species: Application of models from the theory of island biogeography in formulating pest control strategies. J. Appl. Ecol. 18: 773-794.

Sterling, W.L. 1978. Fortuitous biological suppression of the boll weevil by the red imported fire ant. Envir. Entomol. 7: 564-568.

Stern, V.M., R.F. Smith, R. van den Bosch, and K.S. Hagen. 1959. The integration of chemical and biological control of the spotted alfalfa aphid; the integrated control concept. Hilgardia 29: 81-101.

Stevens, L.M., A.L. Steinhauer, and J.R. Coulson. 1975. Suppression of Mexican bean beetle on soybeans with annual inoculative releases of *Pediobius foveolatus.* Envir. Entomol. 4: 497-502.

Stimac, J.L., and R.J. O'Neil. 1985. Integrating influences of natural enemies into models of crop/pest systems. pp. 323-344 in M.A. Hoy and D.C. Herzog, eds. *Biological control in agricultural IPM systems.* Academic Press, New York.

Stinner, R.E., A.P. Gutierrez, and G.D. Butler, Jr. 1974. An algorithm for temperature-dependent growth rate simulation. Canad. Entomol. 106: 519-524.

Stinner, R.E., C.S. Barfield, J.L. Stimac, and L. Dohse. 1983. Dispersal and movement of insect pests. Ann. Rev. Entomol. 28: 319-335.

Stinson, C.S.A. 1983. Effects of insect behaviour on early successional habitats. PhD thesis, University of London.

Strong, D.R., Jr. 1974. Rapid asymptotic species accumulation in phytophagous insect communities: The pests of cacao. Science 185: 1064-1066.

Strong, D.R., Jr. 1979. Boigeographic dynamics of insect host-plant communities. Ann. Rev. Entomol. 24: 89-119.

Strong, D.R. Jr. 1984. Density-vague ecology and liberal population regulation in insects. pp. 313-327 in P.W. Price, C.N. Slobodchikoff, and W.S. Gaud. *A new ecology: Novel approaches to interactive systems.* Wiley, New York. 515 pp.

Strong, D.R., Jr., E.D. McCoy, and J.R. Rey. 1977. Time and the numbers of herbivore species: The pests of sugarcane. Ecology 58: 167-175.

Strong, D.R., J.H. Lawton, and R. Southwood. 1984. *Insects on plants community patterns and mechanisms.* Harvard University Press, Cambridge. 313 pp.

Summerlin, J.W., H.D. Petersen, and R.L. Harris. 1984. Red imported fire ant (Hymenoptera: Formicidae) effects on the horn fly (Diptera: Muscidae) and coprophagous scarabs. Envir. Entomol. 13: 1405-1410.

Summers, C.G. 1976. Population fluctuations of selected arthropods in alfalfa: Influence of two harvesting practices. Envir. Entomol. 5: 103-110.

Summers, C.G., R.E. Garrett, and F.G. Zalom. 1984. New suction device for sampling arthropod populations. J. Econ. Entomol. 77: 817-823.

Suttman, C.E., and G.W. Barrett. 1979. Effects of Sevin on arthropods in an agricultural and an old-field plant community. Ecology 60: 628-641.

Tabashnik, B.E. 1983. Host range evolution: The shift from native legumes to alfalfa by the butterfly *Colias philodice eriphyle.* Evolution 37: 150-162.

Tabashnik, B.E. 1986. Model for managing resistance to fenvalerate in the diamondback moth (Lepidoptera: Plutellidae). J. Econ. Entomol. 79: 1447–1451.

Tabashnik, B.E., and B.A. Croft. 1982. Managing pesticide resistance in crop–arthropod complexes: Interactions between biological and operational factors. Envir. Entomol. 11: 1137–1144.

Takken, W., M.A. Oladunmade, L. Dengwat, H.U. Feldmann, J.A. Onah, S.O. Tenabe, and H.J. Hamann. 1986. The eradication of *Glossina p. palpalis* (Robineau-Desvoidy) (Diptea: Glossinidae) using traps, insecticide-impregnated targets, and the sterile insect technique in central Nigeria. Bull. Entomol. Res. 76: 275–286.

Tallamy, D.W. 1985. Squash beetle feeding behavior: An adaptation against induced cucurbit defenses. Ecology 66: 1574–1579.

Tallamy, D.W., and R.F. Denno. 1982. Life history tradeoffs in *Gargaphia solani* (Hemiptera: Tingidae): The cost of reproduction. Ecology 63: 616–620.

Tamaki, G., R.L. Chauvin, H.R. Moffitt and K.D. Mantey. 1984. Diflubenzuron: Differential toxicity to larvae of the Colorado potato beetle (Coleoptera: Chrysomelidae) and its internal parasite, *Doryphorophaga doryphorae* (Diptera: Tachinidae). Canad. Entomol. 116: 197–202.

Tanada, Y. 1968. The role of viruses in the regulation of the population of the armyworm, *Pseudaletia unipunctata* (Haworth). pp. 25–31 in Proc. Joint U.S.–Japan Seminar on Microbial Control of Insect Pests. U.S.-Japan Comm. Sci. Coop. Panel 8, Fukuoka.

Tauber, M.J., M.A. Hoy, and D.C. Herzog. 1985. Biological control in agricultural IPM systems: A brief overview of the current status and future prospects. pp. 3–9 in M.A. Hoy and D.C. Herzog, eds. *Biological control in agricultural IPM systems.* Academic Press, New York.

Taylor, F. 1979. Convergence to the stable age distribution in populations of insects. Amer. Nat. 113: 511–530.

Taylor, L.R. 1961. Aggregation, variance, and the mean. Nature 189: 732–735.

Taylor, L.R. 1984. Assessing and interpreting the spatial distributions of insect populations. Ann. Rev. Entomol. 29: 321–357.

Taylor, L.R., I.P. Woiwod, and J.N. Perry. 1978. The density dependence of spatial behaviour and the rarity of randomness. J. Anim. Ecol. 47: 383–406.

Taylor, L.R., I.P. Woiwod, and J. N. Perry. 1979. The negative binomial as a dynamic ecological model for aggregation and the density dependence of k. J. Anim. Ecol. 48: 289–304.

Taylor, L.R., R.A. French, I.P. Woiwod, M.J. Dupuch, and J. Nicklen. 1980. Synoptic monitoring for migrant insect pests in Great Britain and western Europe. I. Establishing expected values for species content, population stability and phenology of aphids and moths. Rothamsted Rep. for 1980, pt. 2, pp. 41–104.

Teetes, G.L. 1980. Breeding sorghums resistant to insects. pp. 457–485 in F.G. Maxwell and P.R. Jennings, eds. *Breeding plants resistant to insects.* Wiley, New York.

Thakre, S.K., and S.N. Saxena. 1972. Effect of soil applications of chlorinated insecticides on amino acid composition of maize (*Zea mays*). Plant Soil 37: 415–418.

Thiele, H.-U. 1977. *Carabid beetles in their environments.* Springer-Verlag, New York, 369 pp.

Thompson, L.C. 1985. Insect pests of forests. pp. 509–551 in R.E. Pfadt, ed. *Fundamentals of applied entomology,* 4th ed. Macmillan, New York.

Tingey, W.M., and S.R. Singh. 1980. Environmental factors influencing the magnitude and expression of resistance. pp. 87–113 in F.G. Maxwell and P.R. Jennings, eds. *Breeding plants resistant to insects.* Wiley, New York.

Trumble, J.T. 1982. Within-plant distribution and sampling of aphids (Homoptera: Aphidae) on broccoli in southern California. J. Econ. Entomol. 75: 587–592.

Turnbull, A.C., and D.A. Chant. 1961. The practice and theory of biological control of insects in Canada. Canad. J. Zool. 39: 697–753.

Turpin, F.T., and A.C. York. 1981. Insect management and the pesticide syndrome. Envir. Entomol. 10: 567–572.

Tweeten, L. 1983. The economics of small farms. Science 219: 1037–1041.

United States Department of Agriculture. 1977. *USDA policy on management of pest problems.* Secretary's Memorandum 1929.

University of California. 1981. *Integrated pest management for alfalfa hay.* Univ. Calif. Publ. 4104. 96 pp.

University of California. 1984. *Integrated pest management for cotton in the western region of the United States.* Univ. Calif. Div. Agr. Nat. Res. Publ. 3305. 144 pp.

Uvarov, B.P. 1966. *Grasshoppers and locusts.* Center for Overseas Pest Res., London. 613 pp.

Valentine, W.J., C.M. Newton, and R.L. Talevio. 1976. Compatible systems and decision models for pest management. Envir. Entomol. 5: 891–900.

van den Bosch, R. 1978. *The pesticide conspiracy.* Doubleday, Garden City, NY. 226 pp.

van den Bosch, R., C.F. Legace, and V.M. Stern. 1967. The interrelationship of the aphid, *Acyrthosiphon pisum,* and its parasite, *Aphidius smithi,* in a stable environment. Ecology 48: 993–1000.

van Emden, H.F. 1963. Observations on the effect of flowers on the activity of parasitic Hymenoptera. Entomol. Mon. Mag. 98: 265–270.

van Emden, H.F. 1966. Plant insect relationships and pest control. World Rev. Pest Contr. 5: 115–123.

van Emden, H.F., V.F. Eastop, R.D. Hughes, and M.J. Way. 1969. The ecology of *Myzus persicae.* Ann. Rev. Entomol. 14: 197–270.

van Lenteren, J.C. 1980. Evaluation of control capabilities of natural enemies: Does art have to become science? Neth. J. Zool. 30: 369–381.

Varley, G.C., and G.R. Gradwell. 1960. Key factors in population ecology. J. Anim. Ecol. 29: 399–401.

Varley, G.C., and G.R. Gradwell. 1965. Interpreting winter moth population changes. Proc. XII Int. Congr. Entomol., pp. 377–378.

Varley, G.C., and G.R. Gradwell. 1968. Population models for the winter moth. pp. 132–141 in T.R.E. Southwood, ed. *Insect abundance.* 4th Symp. Roy. Entomol. Soc. Blackwell, London.

Veber, J. 1964. Virulence of an insect virus increased by repeated passages. Entomophaga Mem. Ser. 2: 403–405.

Verhulst, P.F. 1838. Notice sur la loi que la population suit dans son accroissement. Corresp. Math. et Phys. 10: 113–121.

Villavaso, E.J. 1981. Field competitiveness of sterile males released in the boll weevil eradication trial, 1979. J. Econ. Entomol. 74: 373–375.

Vinson, S.B. 1975. Biochemical coevolution between parasitoids and their hosts. pp. 14–48 in P.W. Price, ed. *Evolutionary strategies of parasitic insects and mites.* Plenum, New York.

Vinson, S.B. 1976. Host selection by insect parasitoids. Ann. Rev. Entomol. 21: 109–133.

Volterra, V. 1926. Variazioni e fluttuazioni del numero d'individui in specie animali conviventi. Mem. Accad. Lincei (6) 2: 31–113.

Waage, J.K., and M.P. Hassell. 1982. Parasitoids as biological control agents: A fundamental approach. Parasitol. 84: 241–268.

Wallace, B. 1985. Reflections on some insect pest control procedures. Bull. Entomol. Soc. Amer. 31: 8–13.

Wang, Y., A.P. Gutierrez, G. Oster, and R. Daxl. 1977. A population model for plant growth and development: Coupling cotton–herbivore interactions. Canad. Entomol. 109: 1359–1374.

Ware, G.C. 1978. *The pesticide book*. Freeman, San Francisco. 197 pp.

Ware, G.C. 1983. *Pesticides theory and application*. Freeman, San Francisco. 308 pp.

Washburn, J.O., and H.V. Cornell. 1981. Parasitoids, patches, and phenology: Their possible role in the local extinction of a cynipid gall wasp population. Ecology 62: 1597–1607.

Way, M.J. 1977. Pest and disease status in mixed stands versus monocultures; the relevance of ecosystem stability. pp. 127–138 in J.M. Cherrett and G.R. Sagar, eds. *Origins of pest, parasite, disease and weed problems*. Blackwell, London.

Wedberg, J.L., W.G. Ruesink, E.J. Armbrust, and D.P. Bartell. 1977. *Alfalfa weevil pest management program*. University of Illinois Coop. Ext. Serv. Circ. 1136. 7 pp.

Welch, S.M. 1984. Developments in computer-based IPM delivery systems. Ann. Rev. Entomol. 29: 359–381.

Welch, S.M., B.A. Croft, and M.F. Michels. 1981. Validation of pest management models. Envir. Entomol. 10: 425–432.

Wellington, W.G. 1979. Insect dispersal: A biometeorological perspective. pp. 104–108 in R.L. Rabb and G.G. Kennedy, eds. *Movement of highly mobile insects: Concepts and methodology in research*. North Carolina State University Press, Raleigh.

Weseloh, R.M. 1981. Host location by parasitoids. pp. 79–95 in D.A. Nordlund, R.L. Jones, and W.J. Lewis, eds. *Semiochemicals: Their role in pest control*. Wiley, New York.

Whalon, M.E., and B.A. Croft. 1984. Apple IPM Implementation in North America. Ann. Rev. Entomol. 29: 435–470.

Whalon, M.E., B.A. Croft, and T.M. Mowry. 1982. Introduction and survival of susceptible and pyrethroid-resistant strains of *Amblyseiulus fallacis* (Acari: Phytoseiidae) in a Michigan apple orchard. Envir. Entomol. 11: 1096–1099.

White, G.B. 1979. Can IPM pay its way? New York Food & Life Sci. Quarterly 12 (2): 12–13.

White, L., Jr. 1967. The historical roots of our ecologic crisis. Science 155: 1203–1207.

White, T.C.R. 1976. Weather, food, and plagues of locusts. Oecologia 22: 119–134.

White, T.C.R. 1978. The importance of a relative shortage of food in animal ecology. Oecologia 33: 71–86.

White, T.C.R. 1984. The abundance of invertebrate herbivores in relation to the availability of nitrogen in stressed food plants. Oecologia 63: 90–105.

Whitham, T.G. 1983. Host manipulation of parasites: Within-plant variation as a defense against rapidly evolving pests. pp. 15–41 in R.F. Denno and M.S. McClure, eds. *Variable plants and herbivores in natural and managed systems*. Academic Press, New York.

Whitham, T.G., and S. Mopper. 1985. Chronic herbivory: Impacts on architecture and sex expression of pinyon pine. Science 228: 1089–1091.

Whittaker, R.H., and P.P. Feeny. 1971. Allelochemics: Chemical interactions between species. Science 171: 757–770.

Wiens, J.A. 1977. On competition and variable environments. Amer. Sci. 65: 590–597.

Wigglesworth, V.B. 1959. *The control of growth and form: A study of the epidermal cell in an insect*. Cornell University Press, Ithaca, NY. 140 pp.

Wilbert, H. 1980. Der Einfluss resistanter Pflanzen auf die Populationsdynamik von Schadinsekten. Z. Angew. Entomol. 89: 298–314.

Williams, W.G., G.G. Kennedy, R.T. Yamamoto, J.D. Thacker, and J. Bordner. 1980. 2-tridecanone: A naturally occurring insecticide from the wild tomato *Lycopersicon hirsutum* f. *glabratum*. Science 207: 888–889.

Wilson, C.L., and C.L. Graham, eds. 1983. *Exotic plant pests and North American agriculture*. Academic Press, New York. 522 pp.

Wilson, M.C., and E.J. Armbrust. 1970. Approach to integrated control of the alfalfa weevil. J. Econ. Entomol. 63: 554–557.

Wilson, M.C., D.P. Broersma, and A.W. Provonsha. 1980. *Fundamentals of applied entomology*, 2nd ed. Waveland Press, Prospect Heights, IL. 166 pp. (5 more volumes in series: 2, Livestock and agronomic crops; 3, Vegetable and fruit pests; 4, Ornamental plants; 5, Household and health pests; 6, Forest pests.)

Wilson, R.L., and F.D. Wilson. 1976. Nectariless and glabrous cottons: Effect on pink bollworm in Arizona. J. Econ. Entomol. 69: 623–624.

Wollkind, D.J., A.J. Hastings, and J.A. Logan. 1980. Functional response, numerical response, and stability in arthropod predator–prey systems involving age structure. Res. Pop. Ecol. 22: 323–338.

Wood, D.L. 1980. Approach to research and forest management for western pine beetle control. pp. 417–448 in C.B. Huffaker, ed. *New technology of pest control*. Wiley, New York.

Wood, F.E., W.H. Robinson, S.K. Kraft, and P.A. Zungoli. 1981. Survey of attitudes and knowledge of public housing residents toward cockroaches. Bull. Entomol. Soc. Amer. 27: 9–13.

Wright, L.C., A.A. Berryman, and B.E. Wickman. 1984. Abundance of the fir engraver, *Scolytus ventralis*, and the Douglas-fir beetle, *Dendroctonus pseudotsugae*, following tree defoliation by the Douglas-fir tussock moth, *Orgyia pseudotsugata*. Canad. Entomol. 116: 293–305.

Wyman, J.A., J.L. Libby, and R.K. Chapman. 1977. Cabbage maggot management aided by predictions of adult emergence. J. Econ. Entomol. 70: 327–331.

Yamamoto, I. 1970. Mode of action of pyrethroids, nicotinoids, and rotenoids. Ann. Rev. Entomol. 15: 257–272.

Yeargan, K.V. 1985. Alfalfa: Status and current limits to biological control in the eastern United States. pp. 521–536 in M.A. Hoy and D.C. Herzog, eds. *Biological control in agricultural IPM systems*. Academic Press, New York.

Yoon, J.S., and R.H. Richardson. 1978. Rates and roles of chromosomal and molecular changes in speciation. pp. 129–143 in R.H. Richardson, ed. *The screwworm problem*. University of Texas Press, Austin.

Zinnser, H. 1938. *Rats, lice, and history*. Little, Brown and Co., Boston. 301 pp.

Zungoli, P.A., and W.H. Robinson. 1984. Feasibility of establishing an aesthetic injury level for German cockroach pest management problems. Envir. Entomol. 13: 1453–1458.

Zwolfer, H., M.A. Ghani and V.P. Rao. 1976. Foreign exploration and importation of natural enemies. pp. 189–207 in C.B. Huffaker and P.S. Messenger, eds. *Theory and practice of biological control*. Academic Press, New York.

INDEX

AAIE; *see* Association of Applied Insect Ecologists
Accuracy (in sampling), 36, 43, 48, 49
Acrididae; *see* Locust
Action level, 145
Aedes, 234
Aerosol, 143
Aesthetic injury level, 6, 19, 211, 232–236
 education and, 34
 variation, 19, 27
 urban environment and, 229, 232
Agricultural Experiment Station, 10–12, 145
Agroecosystem, 8, 89, 117, 176, 201, 237
 disruption of, 131, 176
 diversity in, 32, 117
 management, 8, 201
Aldicarb, 4, 137, 143
Aldrin, 139
Alfalfa, 31, 54–56, 183, 211–219
 blotch leafminer, 41, 219
 caterpillar, 31, 180, 218
 insecticide use on, 26, 213, 216, 217
 model, 211–215
 pest complex, 171, 180, 218, 219
 potato leafhopper, 54–56, 215–218
 resistant varieties, 191–194
 sampling, 54–56, 78–80, 216, 217
 seed chalcid, 218
 strip harvesting, 180
 trap crop, 82
Alfalfa weevil, 62, 78–81, 209, 212–217
 biological control, 78–80, 154, 173–177
 diapause, 91
 economic injury level, 213, 216

 Egyptian, 55, 212, 218
 fungal pathogen, 78, 80, 160, 213
 integrated management, 180, 213–215
 life table, 72, 78–80, 88
 parasitoids; *see Bathyplectes; Microctonus; Patasson; Tetrastichus*
 population dynamics, 78–80
 sampling, 49, 56, 209, 216
Allelochemical, 125, 126, 128, 130, 166, 179, 192
Allethrin, 138
Allomone, 166, 187, 189
Amblyseiulus fallacis, 153, 180
American Registry of Professional Entomologists, 14
Animal and Plant Health Inspection Service, 13, 204–206, 211, 214
Anopheles, 6, 30, 165, 234
Ant, 103, 166, 174, 183, 234
 fire, 30, 165, 226
Anthocoridae, 183
Antibiosis, 181, 187, 188, 190
Antixenosis, 188, 193, 225
Aphid, 27, 32, 56, 63, 108, 124, 125, 130, 166; *see also* Green peach aphid; Greenbug
 biological control, 97, 103, 153, 160, 163, 177–183
 biotypes, 27, 89, 190
 blue alfalfa, 219
 cabbage, 64, 124, 193, 219, 220
 corn leaf, 122
 cultural control, 200–202
 dispersion, 37–41, 60, 61

Fruit, 110, 125, 205; *see also* Apple;
  Citrus; Grape; Peach; Pear
  biological control, 185
  economic injury level, 33, 134
  insecticides, 134, 141
  plant resistance, 191, 194
Fruit flies, 168, 198, 205, 206; *see also*
  Mediterranean fruit fly (Medfly)
  Drosophila, 88
  eradication, 168, 205, 206
  sterile-male release, 198, 205
Fumigation, 63, 106, 141, 142, 235
Functional response, 97–99, 101, 176, 184
Fungus, 78, 80, 84, 157, 160, 213

G

Gall, 128, 175, 206, 232
Gambusia, 185
Gargapha solani, 108
Gene flow, 64
Gene-for-gene relationship, 191
Genetic feedback, 110, 131
Genetic manipulations, 195–199
Genetic resistance, 188
Gibberellin, 126
Glasshouse; *see* Greenhouse
Glaucopsyche lygdamus, 126
Glossina, 198
Goldfish, 144
Grain storage, 63, 95, 203
Granule (formulation), 143
Granulosis virus, 159
Grape, 187, 234
  berry moth, 234
  phylloxera, 187, 192
Grasshopper; *see* Locust
Green cloverworm, 46
Green peach aphid, 124, 201, 219, 221
  dispersion, 38–41
  economic injury level, 27, 221
  sampling, 39, 44–47, 60, 61
Greenbug, 32, 89, 189, 190, 193
Greenhouse, 163, 165, 189
  aphids, 160, 165
  biological control, 163, 171, 178
  integrated control, 163
  leaf miners, 33
  whitefly, 128
Ground beetles, 56, 182

Gypsy moth, 12, 31, 91, 131, 228, 230, 231
  biological control, 38, 231
  diapause, 91
  economic injury level, 24, 25, 230, 231
  eradication, 167, 204
  insecticidal control, 135, 164
  pheromone, 167, 168
  sampling, 38
  virus, 158

H

Hatch Act, 10
Hawthorn, 31
Heliothis, 3, 102, 128, 148, 154, 201, 202, 219, 224, 225
  biological control, 179, 181
  corn, 32, 92, 102, 179
  cotton, 3, 5, 150, 201, 202, 224, 225
  cultural control, 201, 202
  dispersal, 92
  NPV, 158, 159, 163
  plant resistance, 32
  resistance to insecticides, 5, 150, 224
  virus, 158, 159, 163
Heptachlor, 139, 213, 235
Hessian fly, 190, 193, 194, 201
Hippodamia convergens, 182
Hirsutella, 158, 160
Honeybee, 34, 88, 136, 137, 149, 156; *see also* Bees
Hormone, 120, 126
Horn fly, 30, 164
Hornworm, 30, 91, 154, 179, 222
House sparrow, 31, 184, 236
Housefly, 64, 91, 152, 203, 234
Hydrogen cyanide, 142
Hypericin, 126
Hyposotor exiguae, 128, 181

I

Ichneumonidae, 56, 60, 182, 228
ICIPE; *see* International Center for Insect Physiology and Ecology
IGR; *see* Insect growth regulator
Immigration, 64, 106, 114, 196
  island biogeography, 106–110